D0745526

FLIGHT

by

Dave Sim and Gerhard

All rights reserved.

Aardvark-Vanaheim Inc.
First printing: (limited signed & numbered) March 1993
Second printing: March 1993
Third printing: December 1993
Fourth printing: November 1996
Fifth printing: December 2001

ISBN 0-919359-13-2

Printed in Windsor, Ontario by
Preney Print & Litho Inc.

PRINTED IN CANADA

Dave's dedication:
to

Deni, Karen, Dawn and Judith

Close but no cigar

Gerhard's dedication:
to Rose,
winner of the 1992
Girlfriend Of The Year Award

Introduction:

Mothers and Daughters (this is the first book of four book series) is the Cerebus story that I have had the longest time to think about. I got the original germ of an idea for the story-line back in 1979, when I first conceived of Cerebus as a three hundred issue series. Essentially, the story concerns the matriarchy which has been founded by Cirin in Upper Felda and the 'daughterarchy' (to coin a word) founded by Astoria in opposition. In the story, the two opposing belief systems are called Cirinism and Kevillism. I gave my speculative side free reign in conceptualizing a matriarchal society which has survived into an industrial age. I found the best means for developing the political structure of an industrial-age matriarchy was to frame an on-going debate in my head between Astoria and Cirin. Whenever I had a spare moment, I would select an element of society and try it out on them. Whichever element was at issue, I had only to try on one voice or the other and its opposite would soon be interrupting the original voice with her objections. I quite enjoyed the process (and still do), because as a male, my opinion was neither desired nor tolerated. Of course, by the time twelve years had gone by, I could've filled the fifty issues of Mothers and Daughters just with an Astoria vs. Cirin debate. That was the genesis of the facing text pages interspersed with the story-line that you will see throughout this and future volumes.

Of course, as with any dramatic device, it becomes difficult not to apply my unusual perceptions to the real world. Why was it that feminists almost universally despised Margaret Thatcher when she was prime minister of England? (answer: Margaret Thatcher was and is a Cirinist, while classic feminist thinking is largely Kevillist in nature). Even the self-description of the two sides in the abortion debate identify themselves along these lines. Pro-Life (all life is sacred, the babies must be safe; Cirinist) and Pro-Choice (I am an individual and I must have control of my own body and life; Kevillist). There are peculiar aberrations as well. Mother Theresa, beaming happily at a newborn child in her hospital in a monstrously over-populated country (the hospital itself in a converted temple of the goddess Kali; goddess of generation and destruction) is obviously a Cirinist without being a mother (except by title). The current head of NOW (whose name escapes me) declared when she was elected to the position that she had a husband and a female lover, but did not consider herself a bi-sexual. Classic Kevillist; individual rights and freedoms above all else, even to the extent of denying an implied definition; I am who I am, not who you think I am., The recent presidential election was less of a race between TweedleClinton and TweedleBush than it was a pitched battle between Cirinist Barbara and Kevillist Hillary; family values versus a co-presidency. Yes, I know the Clintons have a daughter. As long as she doesn't interfere with Hillary's wielding of power, I'm sure she gets a certain amount of attention. 'Quality time' is a phrase invented by Kevillists because their careers don't allow for 'quantity time'.

I could go on for hours.

Just read the book.

And you'll find out why no woman can go out with me for longer that three months.

Dave Sim

Kitchener, Ontario

February 1, 1993

I am a stag; *of seven tines*
I am a flood; *across a plain*
I am a wind; *on a deep lake*
I am a tear; *the Sun lets fall*
I am a hawk; *above the cliff*
I am a thorn; *beneath the nail*
I am a wonder; *among flowers*
I am a wizard; *who but I*
Sets the cool head aflame with smoke?

I am a spear; *that roars for blood*
I am a salmon; *in a pool*
I am a lure; *from paradise*
I am a hill; *where poets walk*
I am a boar; *ruthless and red*
I am a breaker; *threatening doom*
I am a tide; *that drags to death*
I am an infant; *who but I*
peeps from the unhewn dolmen arch?

I am the womb *of every holt*
I am the blaze *on every hill*
I am the queen *of every hive*
I am the shield *for every head*
I am the tomb *of every hope*

mothers
&
daughters

YOU'RE GOING TO NEED HELP GETTING THIS BATCH TO THE FURNACES, CLARINDA...
YES, O CIRIN.

IF YOU'LL FORGIVE A HUMBLE INQUIRY, BLESSED CIRIN ...
HAVE YOU FOUND ANY BOOK WHICH WOULD ASSIST YOU IN THE ONE, TRUE ASCENSION?

THE PAPAL LIBRARY OF THE EASTERN CHURCH, CLARY, IS LIKE A STABLE WHICH HAS NOT BEEN CLEANED IN MANY GENERATIONS.
FIRST, WE GET RID OF THE MANURE ...
THEN, WE ATTEND TO THE HORSES ...

CRING

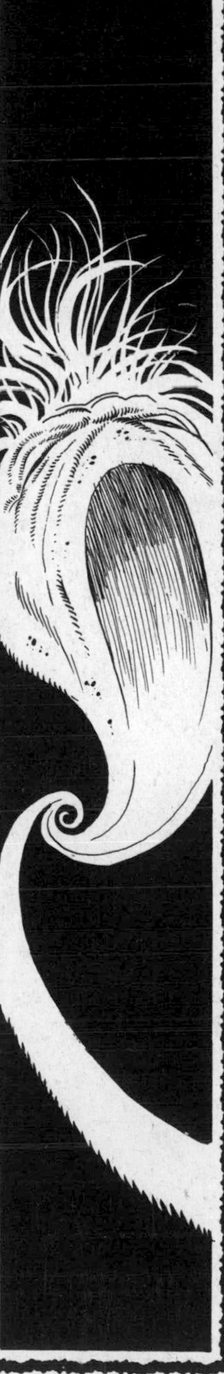

At that moment, beneath a snow-ravaged ravine in far-off Tansubal, the ancient demon, Khem, is the first to feel the coming of the Great Change even as Cirin purges the Papal Library of unwanted works.

Khem's mystic spell was long-broken. It was alone, its captured souls long-departed into the icy skies above a land where winter was eternal and remorseless. No longer did the drained, animated remains of its victims carve hideous representations of the Demon in the soft out-croppings of rock which formed its labyrinthine domain.

Without its spell; without the ability to draw life to it; to drain that life of its vitality and to enslave it, Khem had no purpose; no reason for existence. That awareness dawned, in that moment, in the core of the demon's consciousness and that awareness was black and terrible; like a vacuum so irresistible that Khem's very existence was drawn to it; turning inward upon itself; curling and swirling and disappearing like thin milk into the folds of some vortex; some whirlpool; of icy, dark water.

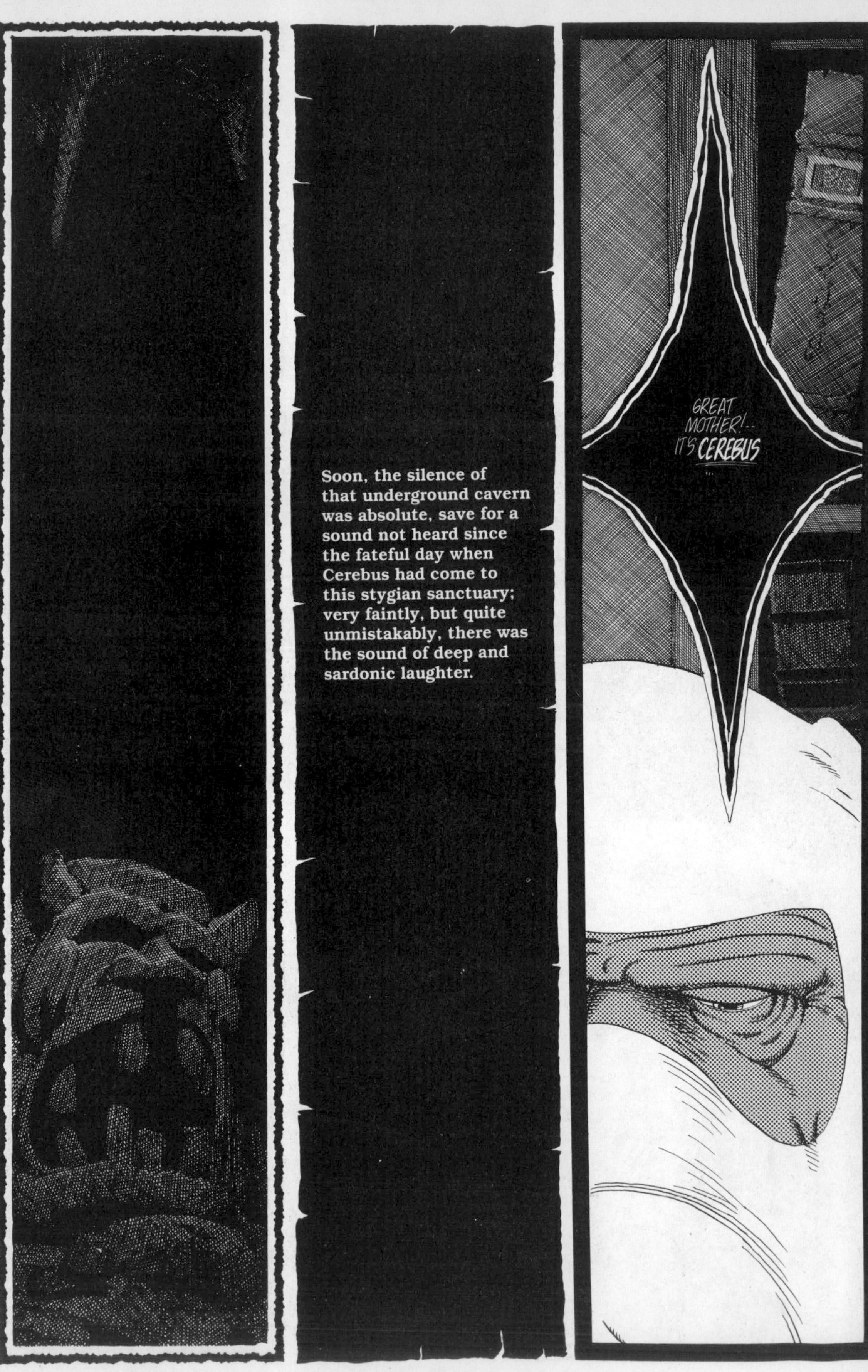
Soon, the silence of that underground cavern was absolute, save for a sound not heard since the fateful day when Cerebus had come to this stygian sanctuary; very faintly, but quite unmistakably, there was the sound of deep and sardonic laughter.
GREAT MOTHER!-- IT'S CEREBUS

OH PLEASE. I'VE ASKED YOU NOT TO DISTURB ME WITH THESE "CEREBUS SIGHTINGS"...
BUT, GREAT CIRIN--
LAST TIME, IF I RECALL CORRECTLY, HE WAS SEEN CLEANING FISH IN THE OLDHAM MARKET
YES, GREAT CIRIN... BUT
AND THE TIME BEFORE THAT HE WAS SEEN PEDDLING FLOWERS ON A STREET CORNER...
YES, O CIRIN--I
THIS TIME, HAS HE AT LEAST FOUND SOME VOCATION COMMENSURATE WITH HIS PREVIOUS STATURE?...
TO ABANDONE ANY NOTION OR IDEA
of the conquest of the upper city came upon
a greate blowe to all of us gathered before
together on the thirde day of our encampment
and spoke most plainly and wisely on the subject
Tarim favourr our course; he saide most

VOCATION? I.. I DON'T ...
AN APPRENTICESHIP WITH ONE OF THE BETTER GUILDS OR A SALARIED POSITION OF SOME ...
GREAT CIRIN..HE HAS MURDERED THREE SOLDIERS --IN COLD BLOOD-- JUST NOW!
WELL, THERE, YOU SEE? SOMEONE'S ...VING A LITTLE JOKE AT YOUR EXPENSE ..
CEREBUS IS...
CEREBUS WAS... ONLY THREE FEET TALL
CRAK
FLIP FLIP FLIP FLI
IT IS ALTOGETHER UNLIKELY THAT ANYONE THREE FEET TALL WOULD MANAGE TO WOUND ONE OF MY SOLDIERS, LET ALONE THREE ...
THAT WILL BE ALL THALISSA ...
BUT, GREAT CIRIN, I...
UNH
YES O'CIRIN

At an altogether different location, a death sits, contemplating his rôle in the larger 'scheme of existence'
~when suddenly

YOU ARE NOT DEATH ...
YOU BEGAN AS A DEMON OF A MOUSE-FRATERNITY IN PRE-SEPRAN TOTEMISTIC ESTARCION, YOU GRADUALLY ROSE IN DIVINE RANK BY FORCE OF ARMS, BLACKMAIL AND FRAUD UNTIL YOU BECAME THE PATRON OF MUSIC, POETRY AND THE ARTS AND FINALLY (IN SOME REGIONS AT LEAST) YOU OUSTED YOUR 'FATHER' TARIM FROM THE SOVEREIGNTY OF THE MULTIVERSE BY IDENTIFYING YOURSELF WITH BELINUS, THE INTELLECTUAL GOD OF LIGHT ...

AS HAPPENS SO OFTEN WITH AN ADVANCEMENT BY NEFARIOUS MEANS, YOU SOON FOUND YOURSELF A VICTIM OF PROFOUND FEELINGS OF PERSECUTION AND WERE SOON SURROUNDED BY ENTITIES WHO SOUGHT TO EXPLOIT THOSE FEARS AND ANXIETIES... OMNIPOTENT AND OMNIPRESENT YOU USED YOUR POWER TO FORGE A NEW IDENTITY FOR YOURSELF; DEATH
WHICH PERMITTED YOU TO VIEW THE VARIOUS FORMS OF WIDE-SCALE CARNAGE AND DESTRUCTION PERPETRATED BY YOUR MORTAL CHILDREN AGAINST THEMSELVES AND EACH OTHER AS A KIND OF PERSONAL TRIUMPH INSTEAD OF A FAILURE AND ABDICATION OF WILL AND INFLUENCE OVER THE MANY WORLDS WITHIN WORLDS YOU FALSELY INHERITED...
HOWEVER!
YOU KNEW THAT AS AN OMNIPOTENT BEING, THE DECEPTION OF THIS NEW IDENTITY WOULD EXTEND (INEVITABLY) TO YOURSELF AND SO YOU CREATED ME...

AND PLANTED THE AWARENESS OF YOUR TRUE IDENTITY IN MY MIND AND TOLD ME ONLY TO REVEAL IT TO YOU IN THE EVENT OF THE APOCALYPSE ARRIVING. THE APOCALYPSE IS NOT ARRIVING (NOT YET ANYWAY), BUT THE TWELVE GEMS SURROUNDING YOU (WHICH AREN'T GEMS AT ALL, BUT LARGE, POWERFUL DISGUISED SUCCUBI) HAVE BEEN DRAINING YOUR STRENGTH AND LIFE ESSENCE FOR THE SEVEN THOUSAND AND SEVEN YEARS, FOUR MONTHS AND EIGHT DAYS SINCE YOU ASCENDED TO THE THRONE OF HEAVEN. SO EFFECTIVE HAVE THEY BEEN THAT YOU HAVE MERE SECONDS LEFT TO YOU...
...AFTER WHICH YOU WILL CEASE TO EXIST AND ALL THAT WILL REMAIN IS A SMALL PILE OF BLACK ROBES AND A LOP-SIDED HOUR-GLASS

ABSOLUTELY NOTHING CAN IMPEDE YOUR IMMINENT DEMISE. THE TWELVE 'GEMS' WILL DISPERSE AT YOUR PASSING AND, SINCE YOU HAVE DONE NOTHING BUT GLOAT OVER DEATH FOR THE LAST FIVE THOUSAND YEARS, NO VOID WILL BE CREATED WHEN YOU GO. IN THE GREAT HIERARCHY OF REALITIES STRETCHING UP TO--AND BEYOND--THE NEARLY LIMITLESS BOUNDARIES OF THE IMAGINATION, YOU WILL BE LONG NOTED AS THE LEAST PRODUCTIVE AND MOST POWERFUL BEING EVER TO HAVE EXISTED...
IT'S NOT MUCH OF A LEGACY, I ADMIT...

BUT IT IS AT LEAST SOMETHING.

POIT:

Well
Fuck
me.

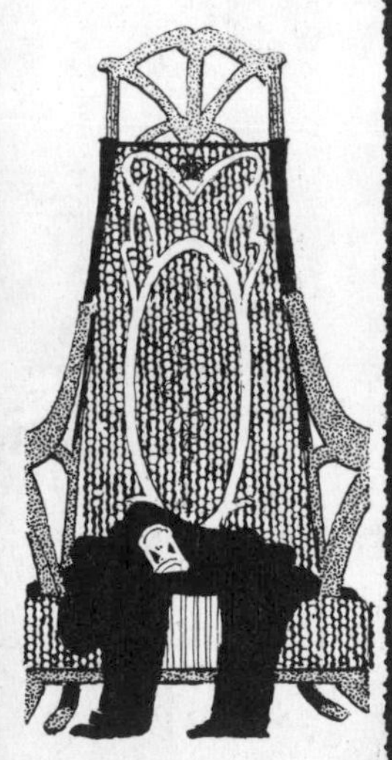

KILL
ILL HIM
HIM KILL
ILL
HIM
KILL
KILL
HIM
KILL
ILL
HIM
KILL
HIM
KILL
HIM
KILL
HIM
KILL
HIM
KI

WHAT DOES SHE MEAN SHE REFUSES TO MEET WITH MY COMMODITIES EXCHANGE POLICY DELEGATION? ...
THEIR RULE IS VERY STRICT LORD JULIUS
NO INTERACTION WITH ANY TRADE GROUP WITH LESS THAN FIFTY PER CENT FEMALE REPRESENTATION ...
FIFTY PER CENT, EH?
YES, LORD JULIUS
HOW MANY FEMALE COMMODITIES EXCHANGE DELEGATES DO WE HAVE?
NONE, LORD JULIUS

HMM.
WELL THAT SHOULDN'T BE TOO MUCH OF A PROBLEM
WE'LL JUST TRAIN SOME
YES, LORD JULIUS
SO WHAT DOES A COMMODITIES EXCHANGE POLICY DELEGATE DO EXACTLY?
THERE'S AN INTERIM REPORT DUE ON THAT FROM THE COMMITEE ON POLICY AND COMMODITY DEVELOPMENT AND EXCHANGE
OH?
AND HOW'S THAT COMING ALONG?
AS IF I DIDN'T KNOW
THE FINAL DRAUGHT HAS BEEN REFERRED TO THE "STEERING" COMMITEE ON PUBLIC POLICY AND EXCHANGE
FLIP FLIP
"STEERING" COMMITEE.
YES LORD JULIUS
WELL, I SUPPOSE I'LL HAVE TO TALK TO THEM

HOW MANY PEOPLE ARE THERE ON THE "STEERING" COMMITEE?
SEVENTY-TWO, LORD JULIUS.
SEVENTY-TWO PEOPLE ON THE "STEERING" COMMITEE ?
YES, LORD JULIUS.
TAKE A LETTER; TO THE CHAIR OF THE "STEERING" COMMITEE
"DEAR CHAIR"
" I HAVE JUST BEEN INFORMED THAT SEVENTY TWO PEOPLE SERVE UNDER YOU ON THE "STEERING" COMMITEE ON PUBLIC POLICY AND EXCHANGE"
" IT IS MY CONSIDERED OPINION THAT SEVENTY-TWO PEOPLE UNDER A CHAIR VIOLATES SEVERAL ARTICLES OF THE PALNAN FIRE CODE"
"NOT TO MENTION THE HEALTH AND SAFETY CHARTER OF 1408 AND MOST PEOPLE'S IDEA OF A GOOD TIME"

"THEREFORE I AM INSTRUCTING YOU THAT AS OF THIS DATE, ALL MEMBERS SERVING ON THE "STEERING" COMMITEE (WITH THE EXCEPTION OF YOURSELF) ARE TO HAVE THEIR TENURE REVOKED"
"YOU MAY INFORM THE MEMBERS OF THIS AND ADVISE THEM THAT THERE ARE (ODDLY ENOUGH) SEVENTY-ONE POSITIONS CURRENTLY AVAILABLE ON THE NEWLY-FORMED "EVERYBODY GET OUT AND PUSH" COMMITEE
"YOURS VERY SINCERELY" ETCETERA ETCETERA

FAR TO THE NORTH-WEST

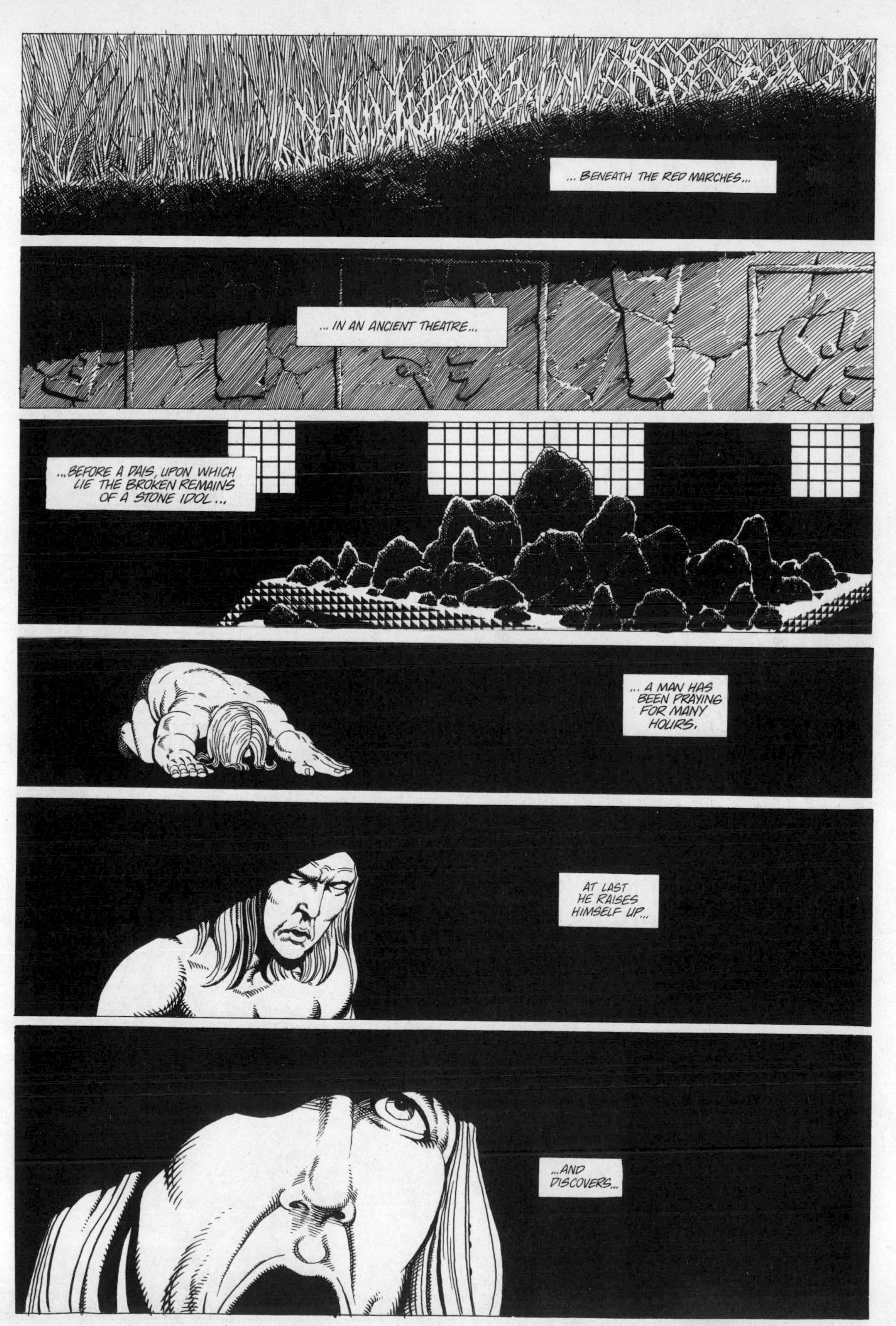
... BENEATH THE RED MARCHES...
... IN AN ANCIENT THEATRE...
...BEFORE A DAIS, UPON WHICH LIE THE BROKEN REMAINS OF A STONE IDOL...
... A MAN HAS BEEN PRAYING FOR MANY HOURS.
AT LAST HE RAISES HIMSELF UP...
...AND DISCOVERS...

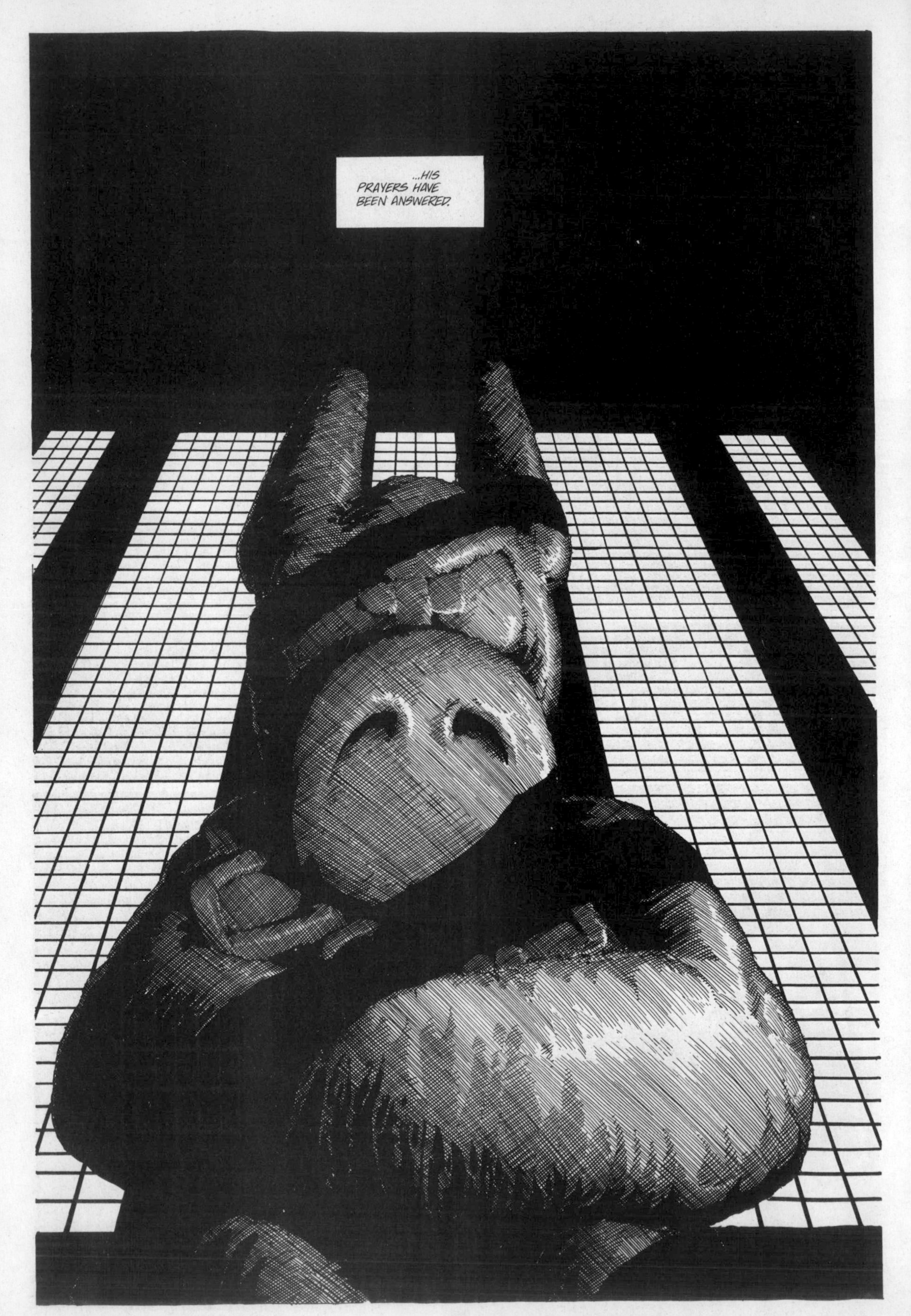
...HIS PRAYERS HAVE BEEN ANSWERED.

KILL HIM
KILL HIM
KIL' HIM.
KILL HIM
KILL HIM

KANG
KILL HIM
KILL HIM
KILL HIM
KILL HIM
KILL HIM
KILL HIM
IT'S THE POPE.
IT'S WHO?

IT'S THE POPE! IT'S CEREBUS ...
DON'T BE RIDICULOUS... CEREBUS DIED WHEN THE ILLUSION-TOWER VANISHED...
EVERYONE KNOWS THAT...

YOU KNOW ANYONE ELSE WHO LOOKS LIKE THAT?

OH!

TOLD YOU IT WAS HIM

HE'S KILLING ALL THE CIRINISTS...
HE'S BEEN RESURRECTED-- THAT'S WHAT!!
BY TARIM!

AND HE'S COME BACK TO REDEEM IEST...
JUST LIKE THE PROPHECIES SAID

YOU UP THERE! GET AWAY FROM THAT WINDOW OR I'LL CUT YOUR EYES OUT!

BETTER DO AS SHE...
YOU DON'T SCARE ME COW!

BRYAN!
CEREBUS HAS COME BACK TO REDEEM US!

HE'LL SORT YOU OUT!
BRYAN DON'T

HAIL CEREBUS!!
YES.
YES!
BRYAN

HAIL CEREBUS!!

HAIL CEREBUS
HAIL CEREBUS
HAIL
KILL HIM
KILL HIM
PLEASE PLEASE LISTEN
KILL HIM
KILL HIM
KILL HIM
KILL HIM
KILL HIM
KILL HIM
KILL HIM
LISTEN
LISTEN
LISTEN
LISTEN LISTEN
LISTEN
LISTEN
WE'VE GOT TO ALL TAKE HIM AT ONCE
AYE
ENCIRCLE HIM
STEADY
CAREFULLY
CAREFULLY
HE'S VERY DANGEROUS USE YOUR HEAD
SHE'S RIGHT EASY DOES IT
DON'T LET HIS SIZE FOOL YOU

HAIL CEREBUS
HAIL CEREBUS CEREBUS
HAIL CEREBUS
CAREFUL.
CAREFUL.
!

HAIL CEREBUS
KILL HIM!
KILL HIM
KILL HIM
KILL HIM

BUS
HAIL CEREBUS
HAIL CEREBUS
HAIL CEREBUS
HAIL CEREBUS
GO HOME FAT, UGLY BITCHES
HAIL THE TRUE POPE
WHAT A MESS! WHAT AN ABSOLUTE MESS!
HOW MANY DEAD?
SIX--ER-- SEVEN-- ER EIGHT, MA'AM
WHAT A MESS ...
AND HAS BLESSED CIRIN NOTHING TO SAY ABOUT ALL THIS?
SHE TOLD US NOT TO BOTHER HER WITH ANY MORE "CEREBUS SIGHTINGS" ...
SHE...?

...OF ALL THE...
CRUN
UNH!
THUMP
WHO-?
WHO? IT'S NORMINA SWARTSKOF--THE ONE WHO CAPTURED THIS PENIS OF A TOWN FOR YOU

LOOK AT THIS MESS
YES GENERAL
WHAT CAN I DO FOR YOU?
JUST LOOK
I CAN ONLY HOPE--FOR YOUR SAKE--THAT THIS IS

WHO--
WHO COULD HAVE DONE SUCH A--
LOOK AROUND
YOU'LL SEE HIM
BUT WHO?!
HOW WILL I--
NO

NO
At that moment, far to the west, in the murky depths of the Feld River, the Black Blossom Lotus rises from the muck and the effluvium of the river bed where it has lain for more than three years. As it does so, a pure white sliver (a feather perhaps, or a leaf) extrudes itself from the Blossom's heart and passes, without obstruction, through the bubble of glass which encloses the night-shaded flower . . .
In the next instant, the Black Blossom Lotus and its covering have vanished, leaving only the thin, brilliant shape hovering in the dark waters, seemingly unaffected by the movement of the surrounding river . . .

Several miles away, two gold coins, minted more than fourteen centuries before, rise, as well, from the bed of that same river; three years worth of encrustations falling from them like raindrops from polished marble . . .
The two coins begin a slow revolution around each other, shining ever more brightly as their relative speed increases. They, too seem immune to the eddies and currents . . .
HAIL CEREBUS
KILL HIM

HAIL CEREBUS
HAIL CEREBUS
HAIL CEREBUS
KILL HIM
KILL HIM
KILL HIM
KILL HIM

HAIL CEREBUS
HAIL CEREBUS
KILL HIM
KILL HIM
KILL HIM
KILL HIM

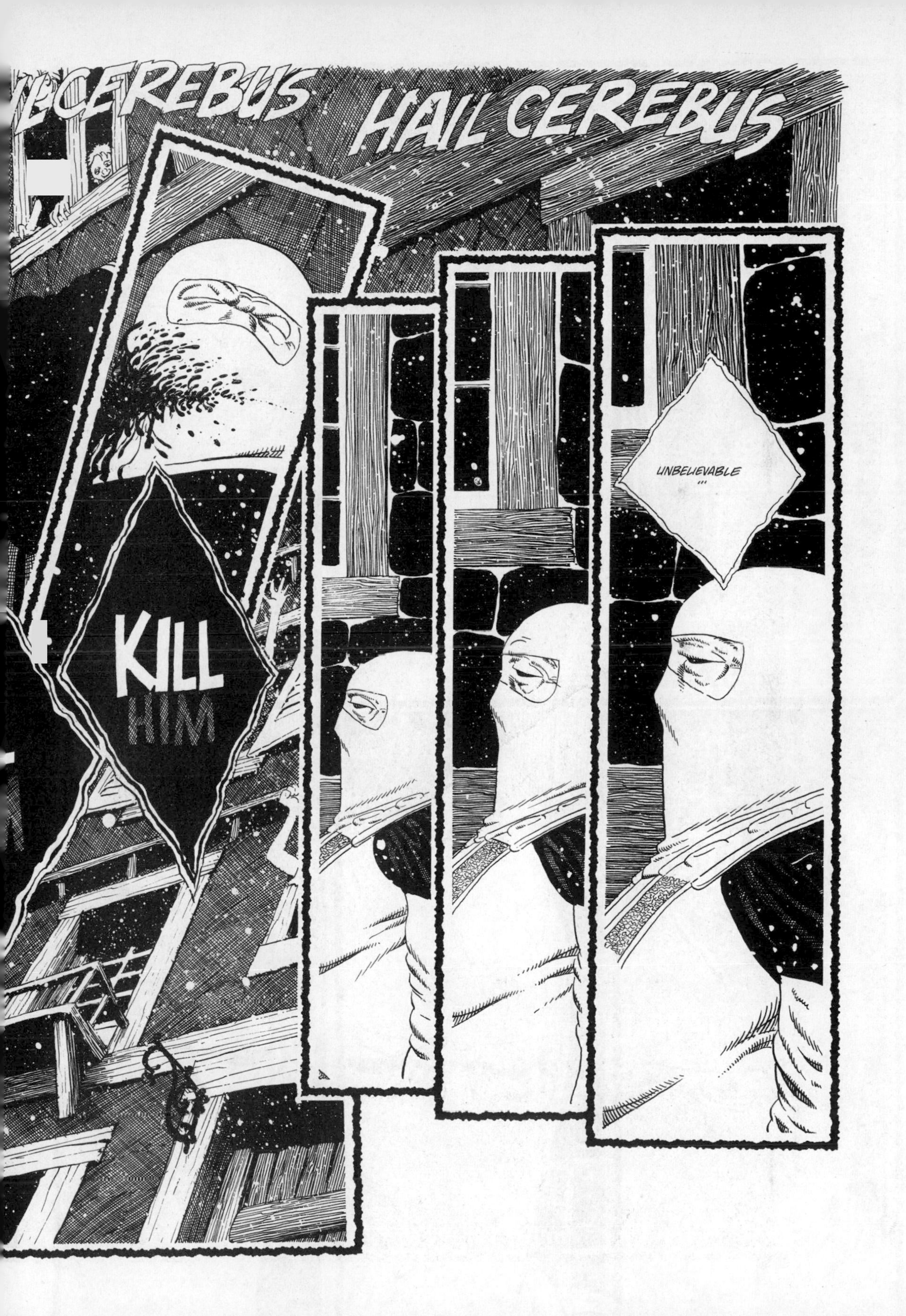

CEREBUS
HAIL CEREBUS
KILL
HIM
UNBELIEVABLE
...

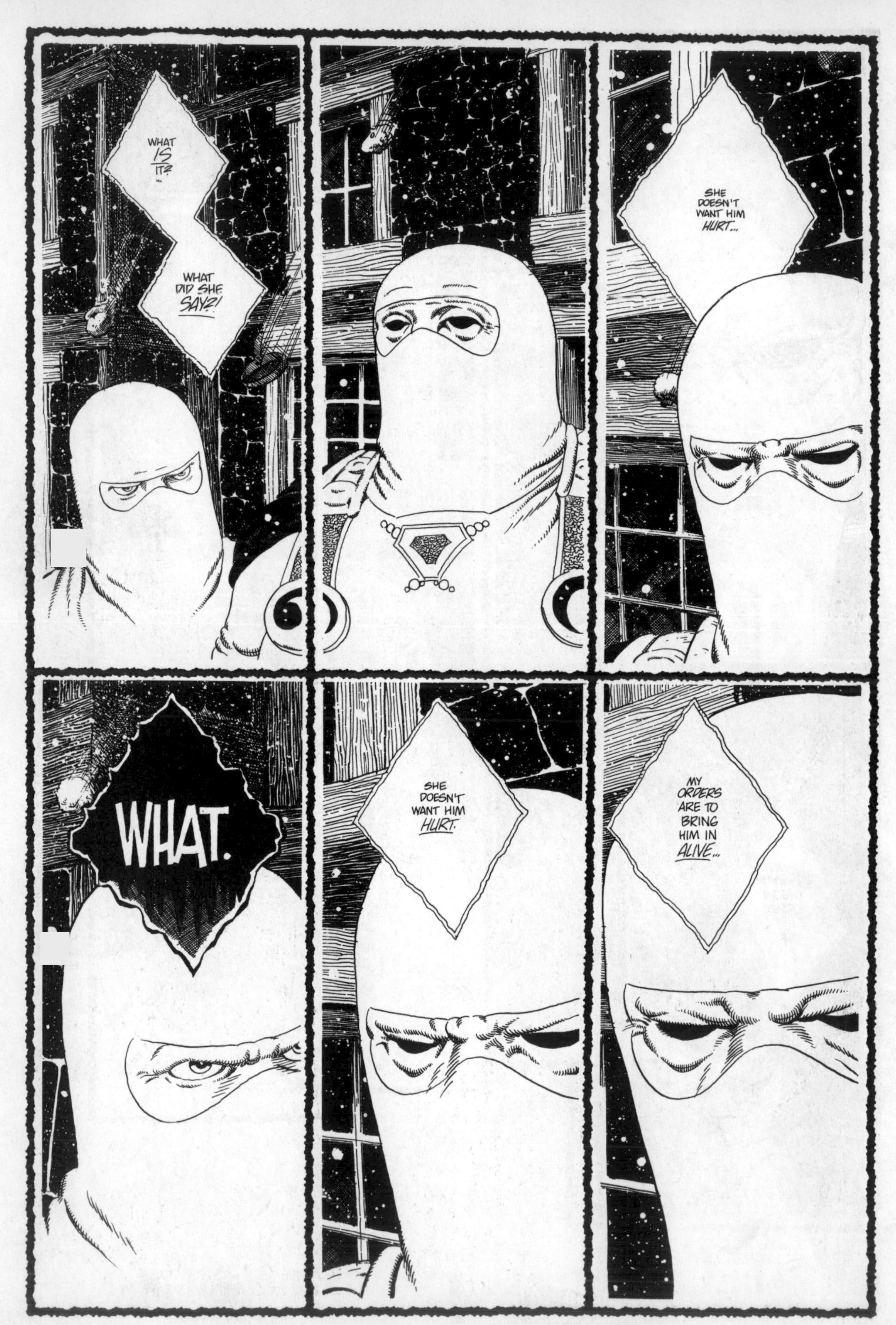
WHAT IS IT? ...
WHAT DID SHE SAY?!
SHE DOESN'T WANT HIM HURT...
WHAT.
SHE DOESN'T WANT HIM HURT.
MY ORDERS ARE TO BRING HIM IN ALIVE...

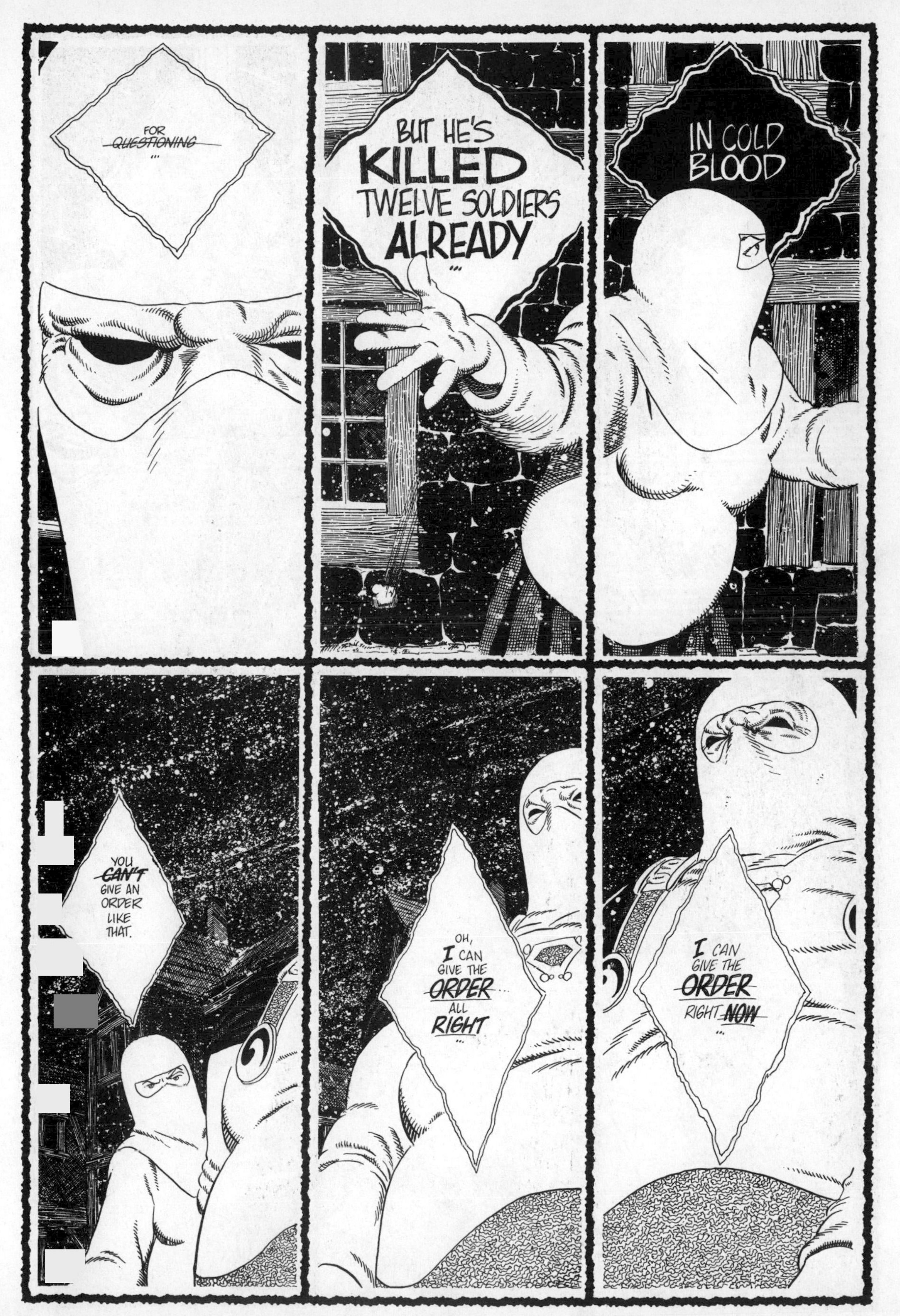
FOR QUESTIONING ...
BUT HE'S KILLED TWELVE SOLDIERS ALREADY ...
IN COLD BLOOD
YOU CAN'T GIVE AN ORDER LIKE THAT.
OH, I CAN GIVE THE ORDER ... ALL RIGHT ...
I CAN GIVE THE ORDER RIGHT NOW ...

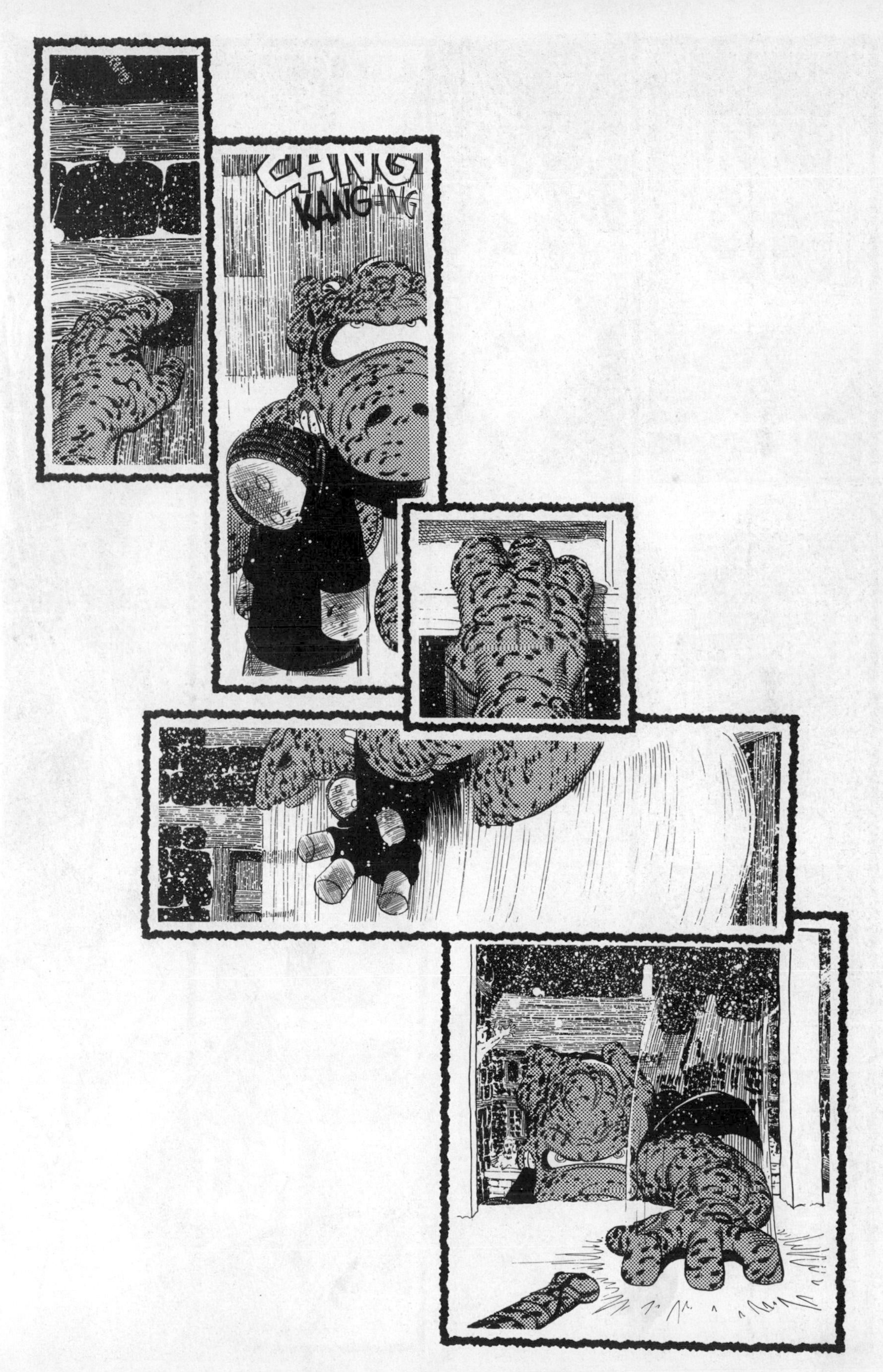
CANG
KANG-ANG

HAIL CEREBUS
HAIL CEREBUS
HAIL CEREB
CLIMBING, EH?
GOOD GOOD!
GET A DOZEN SOLDIERS ON THAT ROOF IMMEDIATELY ...
AND ANOTHER DOZEN UNDERNEATH HIM IN CASE HE DECIDES TO BACK-TRACK
YES, GENERAL ...
KILL HIM
KILL HIM
KILL HIM

HAIL CEREBUS
HAIL CEREBUS
HAIL CEREBUS
THIS SHOULDN'T TAKE LONG
THIS SHOULDN'T TAKE LONG AT ALL ...
KILL HIM
KILL HIM

CITIZENS OF IEST! ....

MOST HOLY
HAS RETURNED

WHAT IN THE...

HURRAHHH

AND BRINGS TO YOU
THE SACRED WORD

OF THE LIVING TARIM

VENGEANCE

FOR THOSE WHO WOULD ENSLAVE IEST!!
FOR THOSE WHO DEFILE THIS SACRED CITY!!
FOR THOSE WHO HAVE SPILLED IESTAN BLOOD!!
THERE CAN BE ONLY ONE PRICE WHICH WILL SATISFY THE LIVING TARIM!

The two coins
accelerate in
their orbit,
glowing ever-
more-brightly;
radiating,
glistening
with white
brilliance

DEATH

DEATH!

DEATH!

DEATH!

DEATH!

SEIZE A WEAPON

A SWORD!

A KNIFE!

AN AXE!

WHATEVER YOU CAN *FIND!*

AYE.

BRYAN?

MY JACKET ...
SHE'S-A LOOK FUNNY
mm.
MAYBE YOU SHOULD SWITCH TAILORS...

GO OUT INTO THE STREETS!
BRYAN! NO!

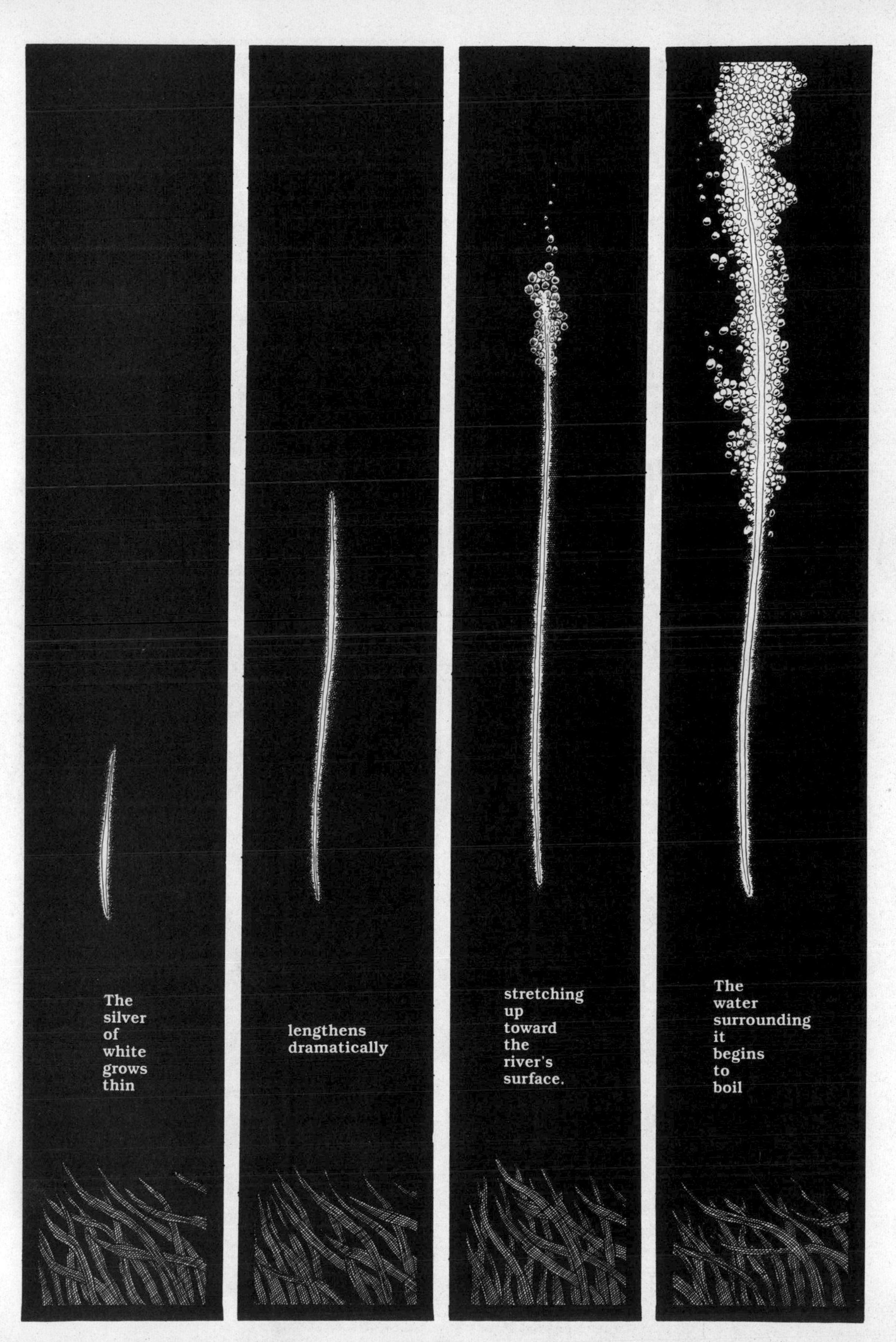
The
silver
of
white
grows
thin
lengthens
dramatically
stretching
up
toward
the
river's
surface.
The
water
surrounding
it
begins
to
boil

TAKE
BACK
YOUR
CITY!

NO NO NO!
MY JACKET
SHE LOOKS
A FINE...
THE WAY Y'U
PAINT A MY
JACKET...
SHE
LOOKS
A FUNNY
HMMM.
YOU MEAN
THE COLOUR
OR THE
...
SETH!

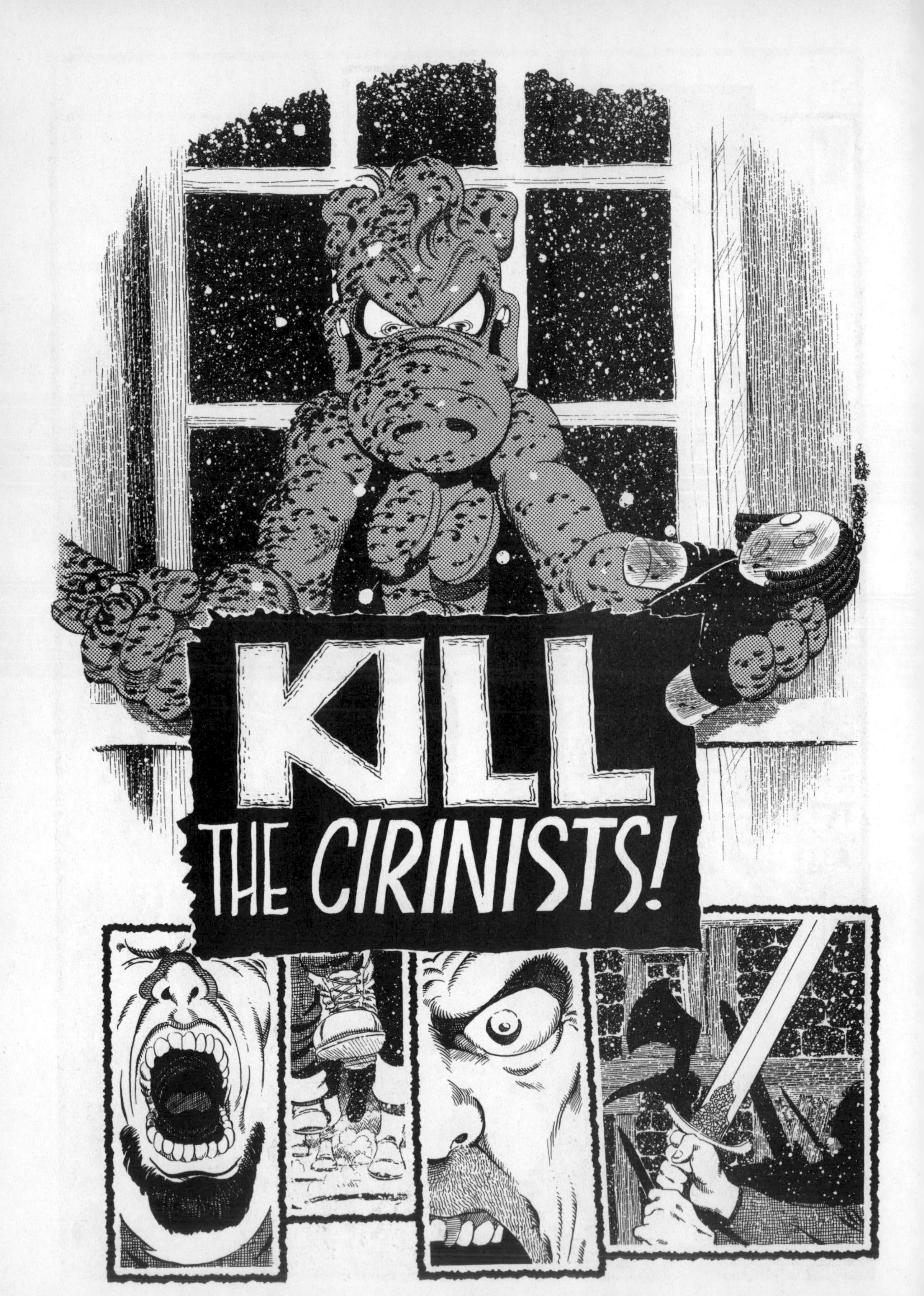
KILL
THE CIRINISTS!

Dear Lord Bradford~
SKTCHA SKTCH
Regarding your letter of
DIP
SKTCHA SKTCHA
CLARINDA!
YES, MRS. THATCHER?

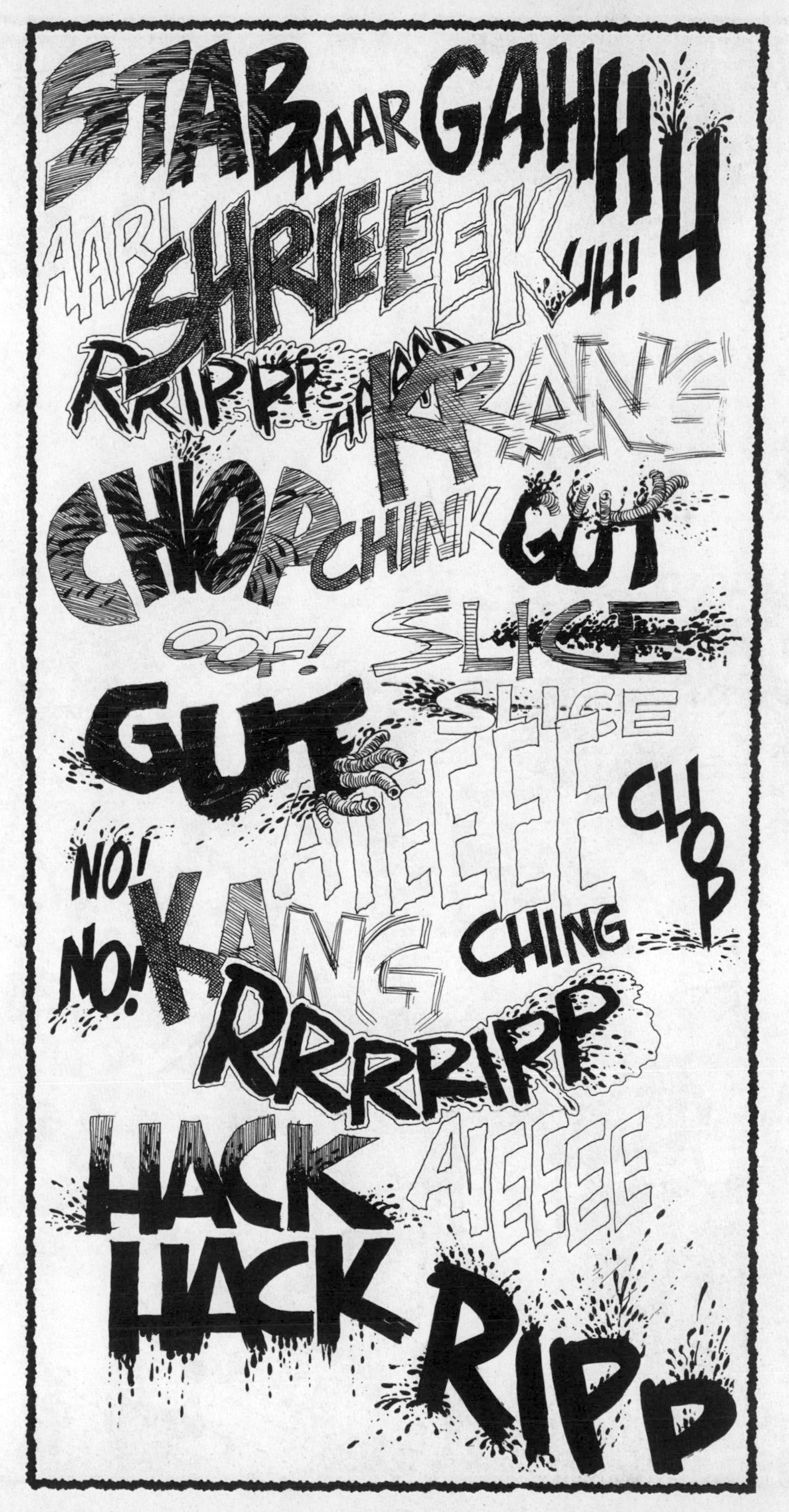

STAB
AAAR GAHHHH
SHRIEEEK
UH!
RRIPPP
CHOP
CHINK
GUT
OOF!
SLICE
GUT
SLICE
AIEEEE
CHOP
NO!
NO!
KANG
CHING
RRRRRIPP
HACK
AIEEEE
HACK
RIPPP

YOU'VE NIBBLED AT MY TEAT LONG ENOUGH YOU LITTLE SHIT

CROSSBOW.

YES GENERAL

CAN YOU HIT THE LITTLE BASTARD FROM HERE ?

MM
WHAT PART OF HIM ?
HIS HEART ...

I WANT A CLEAN KILL

I THOUGHT YOU WANTED SOMETHING DIFFICULT --LIKE HIS THUMB OR A TOE-NAIL

SLAM
SLAM
WATCH THIS ...
SLAM

SMASH!
!
?

HE WAS RIGHT THERE!
HMMMM.
IT WAS HUGE, CLARINDA ...
GRRROWING OUT OF THE TOP OF THE MOUNTAIN ...

AND IT WAS SHAPED LIKE A... A...
IT LOOKED EXACTLY LIKE A...
IT WAS A GREAT BIG
THAT WILL BE ALL CLARINDA
THANK YOU
?
YES MRS. THATCHER

ZZZZ
CIRIN.
YES, GENERAL. YOU'VE CAPTURED HIM?
HE'S UNHURT?
HE DISAPPEARED.
disappeared
...

"DISAPPEARED"?
WHAT DO YOU MEAN "DISAPPEARED"?
ONE OF MY SOLDIERS WAS JUST ABOUT TO ...
... TO CAPTURE HIM...
AND HE VANISHED
INTO THIN AIR
IT'S TIME

THE ILLUSIONISTS, THEN
SUENTELUS PO MUST BE BEHIND IT.
I'VE GOT TWELVE DEAD SOLDIERS DOWN HERE
WHAT SORT OF "ILLUSION" DO YOU CALL THAT?
CHECK THE BODIES
MAKE SURE THEY'RE REAL...

I REMEMBER NOW!
I'M THE MERCHANT!
THE COCKROACH!
CAPTAIN COCKROACH!
THE MOON ROACH!
WOLVERROACH!
SGT. PRESTON!
THE PURELY PLATONIC PRIEST ROACH!!
THE SECRET SACRED WARS ROACH
normalroach...
I WAS USED BY WEISSHAUPT
BY ASTORIA!
BY CHARLES X. CLAREMONT
I REMEMBER!!
IREMEMBER I REMEMBER I REMEMBER I REMEMBER I REMEMBER
EPOP THE ANTI-POPE!!
MICHELLE!
THE FINAL ASCENSION
BUNKY THE ALBINO!!
BETH, DEAR, WE'RE READy FOR YOU
HE'S HERE ...SOME-WHERE

POWERS (PRIEST)
WEISSHAUPT (KING)
CEREBUS BECOMES PRIME MINISTER AND THEN POPE (KING AND THEN PRIEST)
JULIUS (KING)
CEREBUS BEATS JULIUS' GOAT (KING BEATS KING'S KING)
ASTORIA KILLS LION... (PRIESTESS BEATS PRIEST)
ASTORIA (PRIESTESS)
CIRIN (PRIESTESS)
CIRIN INVADES AND CONQUERS IEST (PRIESTESS BECOMES QUEEN)
(PRIESTESS BEATS QUEEN)
CEREBUS CAPTURES ASTORIA (PRIEST BEATS PRIESTESS)
(MAGICIAN BEATS PRIESTESS)
PRIEST ECOMES AGICIAN.
CEREBUS BECOMES MAGICIAN...
ASTORIA (PRIESTESS) IMPRISONED CIRIN BECOMES (QUEEN OVER PRIESTESS) CIRIN OVER JULIUS (QUEEN OVER KING) (OVER) (POWERS) (PRIEST) ...
MAGICIAN ...
DISAPPEARS

NO! NOT 'DISAPPEARS'
'ASCENDS'
AND IT CAME TO PASS IN THE VALLEY OF THE FELD RIVER, IN THE LAND OF ESTARCION IN THE CITY-STATE OF IEST, THIRD STREET ON THE LEFT YOU CAN'T MISS IT THAT CEREBUS ASCENDED INTO VANAHEIM, RISING UP UP UP ON THE BLACK TOWER INTO THE HEAVENS WHERE HE MET THE LIVING TARIM WHO SPAKE UNTO HIM SAYING "WOE BETIDE THE DAUGHTERS OF SAPPHO-- WOE UNTO THEIR DAUGHTERS' DAUGHTERS WHO HAVE RISEN UP AND KNOCKED THE COSMIC APPLECART ALL KERFLOOEY-- WOE, WOE A THOUSAND TIMES WOE! YOU WILL RETURN, CEREBUS MY ONLY MISBEGOTTEN SON AND SPAKE UNTO THE PEOPLE AND KEEP SPAKING UNTIL THEY GRABBETH A CLUE ABOUT THE WORD OF THE LIVING TARIM WHICH IS...

VENGEANCE
THE BODIES ARE REAL
NO QUESTION
IN THAT CASE, GENERAL, WE HAVE QUITE A PROBLEM ON OUR HANDS
VENGEANCE CRIES HIS SWORD

VENGEANCE CRIES MISSY
EPIPHANY
EPIPHANY
EPIPHANY
EPIPHANY
BOUNCE
BOUNCE
HOW MANY PEOPLE SAW HIM ...
... SAW THE ILLUSION?
DOZENS.
HUNDREDS.

MMMM.
TIME TO...
WASH MY HAIR!
LEAP!
SLAM
QUARANTINE THE ENTIRE AREA...
TWO MILE RADIUS
NO ONE GETS OUT NO ONE GETS IN...

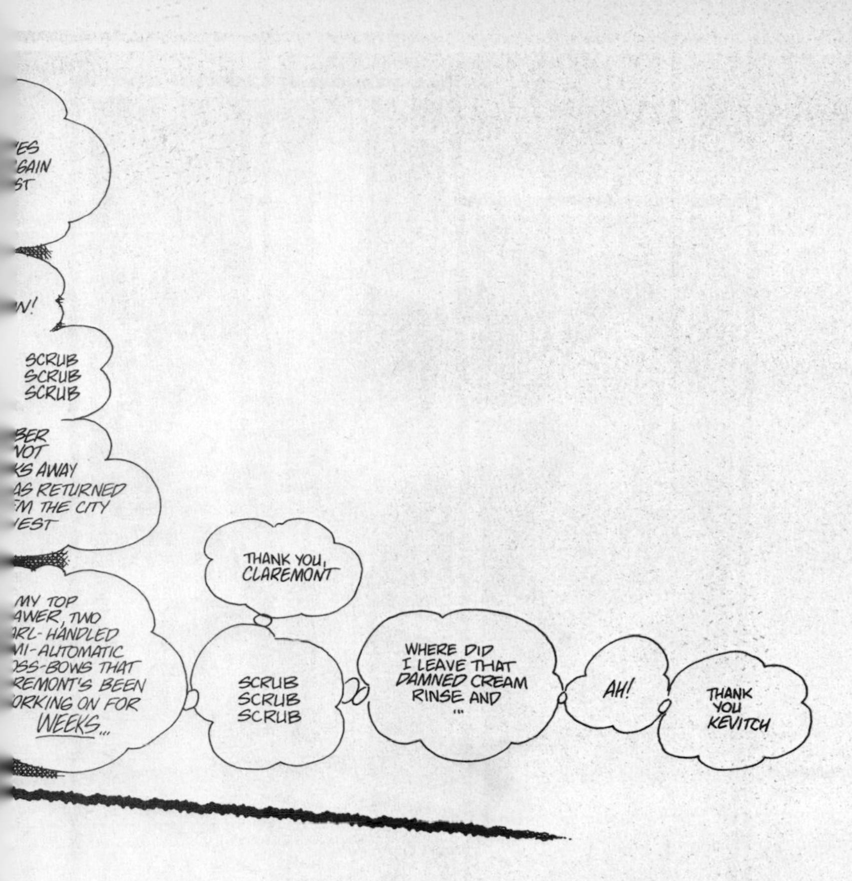
SCRUB SCRUB SCRUB
THANK YOU, CLAREMONT
SCRUB SCRUB SCRUB
WHERE DID I LEAVE THAT DAMNED CREAM RINSE AND ...
AH!
THANK YOU KEVITCH

RELOCATE ALL THE MOTHERS AND CHILDREN UNDER FIVE TO HEALTH CENTERS IN UPPER FELDA
EXECUTE EVERYONE ELSE

RUB RUB RUB
COMB COMB C...
TEASE TEASE TEASE
COMB COMB COMB COMB COMB
TWO BLOCKS AWAY CEREBUS HAS CALLED FOR THE EXECUTION OF THE HATED CIRINISTS... MEN ARE POURING INTO THE STREETS WITH HOME-MADE IMPLEMENTS STRIKING THEM DOWN LEFT AND RIGHT AND RIGHT AND LEFT AND
NOW WHERE DID I PUT THAT STYLING GEL?
AH
THANK YOU KEVITCH ...
HE'S GONE.
IT WAS HIM THOUGH, WASN'T IT?

EXECUTE ...?
YOU DON'T REALLY THINK WE'LL FIND EVERYONE WHO SAW HIM DO YOU?
THE ODDS ARE ...
THE ODDS ARE GETTING WORSE EVERY SECOND YOU STAND AROUND TELLING ME WHAT THE ODDS ARE, GENERAL ...

WHAT ARE WE GOING TO DO?

YOU'RE THE BOSS.
THE GREAT CHANGE HAS BEGUN--A CITY-WIDE POPULAR UPRISING! THE STREETS WILL RUN RED WITH THE BLOOD OF THE UNBELIEVERS! BY SUNSET WE WILL HAVE BEGUN THE BATTLE FOR CONTROL OF THE UPPER CITY! BY MID-NIGHT CIRIN WILL BE HANGING FROM A LAMP-POST AND CEREBUS WILL ONCE MORE ASCEND TO THE PAPACY OF THE EASTERN CHURCH AND I WILL BE HIS STRONG RIGHT ARM, MY REWARD FOR VALOR IN THE GREAT PATRIOTIC WAR AND MY INSTALLATION IN SUCH A POSITION OF POWER AND AUTHORITY WILL SIGNAL THE FIRST GREAT VICTORY OF THE SIX-FOOT TELEPATHIC COCKROACHES PING PING PING
TOOTHPASTE TOOTHPASTE
WUMBSHUMSH SHUMSHEEMA SHOOMSHEM SHESHMAMASH MEEESH...
GRGLRGRGL
PTOO

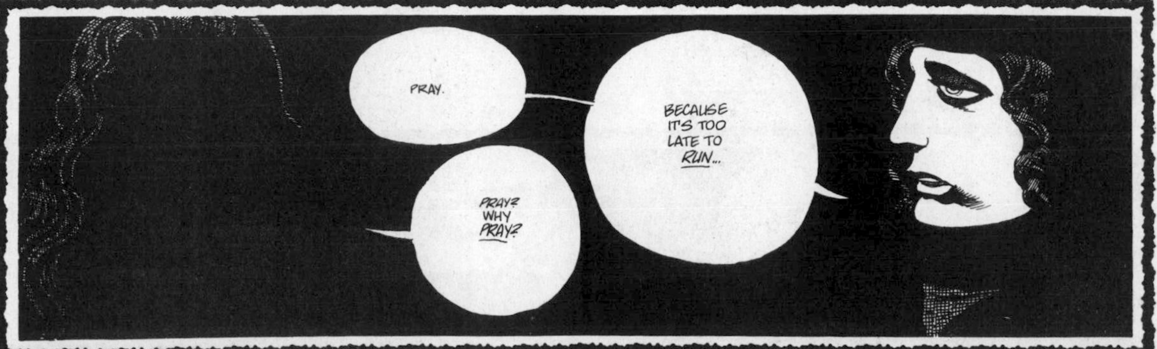
PRAY.
PRAY? WHY PRAY?
BECAUSE IT'S TOO LATE TO RUN...

MUST MOVE AT NEARLY SUPERHUMAN SPEED PAUSING ONLY LONG ENOUGH TO SNATCH UP THE PEARL-HANDLED SEMI-AUTOMATIC CROSSBOWS AND CROSSBOW BOLT BELTS FOR THE GREAT CHANGE HAS ALREADY BEGUN-- THE SECOND COUSIN OF ALL BATTLES RAGES NEARBY AND ONLY THE NEWLY UNIFIED MULTI-ROACH CAN TIP THE SCALES IN FAVOUR OF THE FORCES OF THE ONE TRUE POPE!

TOO LATE.

CAAAREEEEEEEEENING AROUND
CORNERS, DOWN HALLWAYS WITH CHARACTERISTIC GRACE AND POWER -- HURTLING DOWN STAIRWAYS AND OUT INTO THE STREET CASTING OFF CLOTHING SO THAT IT SEEMS FOR THE MOMENT THAT THE LATEST ROACH INCARNATION WILL BE EXHIBITIONIST ROACH. IT IS A MATTER OF A FEW INSTANTS BEFORE I REMEMBER THE COSTUME SERGEANT PRESTON HAS BEEN WORKING ON FOR SEVERAL WEEKS WHICH I'M WEARING UNDER MY normal roach DISGUISE... THANK YOU SERGEANT PRESTON
HOTEL
?
THE DIE IS CAST!
THE TIME IS NIGH ...
LEAP
THE MOMENT IS AT HAND!
LEAP
NOW IT BEGINS
LEAP
THE COLLECTOR'S EDITION DOUBLE BAG FIRST APPEARANCE OF

PUNISHERORCH
ALL RIGHT YOU STINKING LEZBOS

LET'S
...

...boogie?

BENEATH THE RED MARCHES, THE ANCIENT THEATRE SOON FILLS WITH REPRESENTATIVES OF EVERY CLAN...

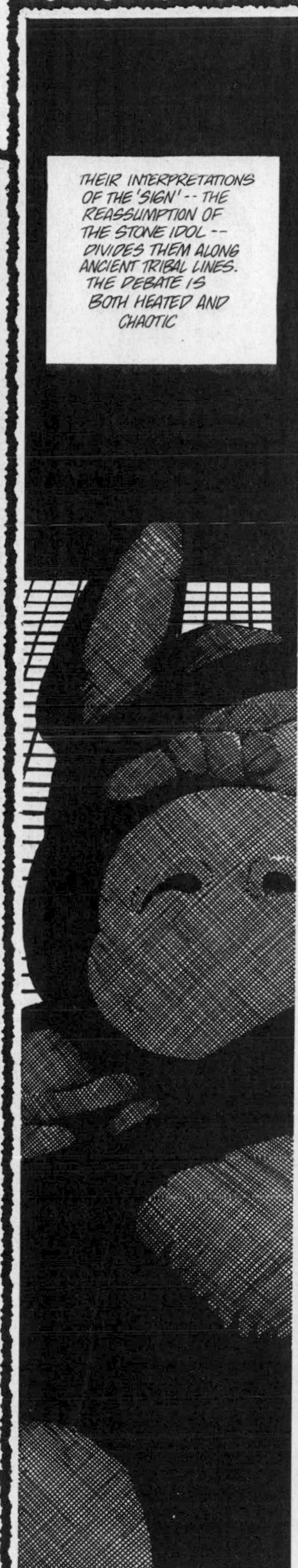
THEIR INTERPRETATIONS OF THE 'SIGN' -- THE REASSUMPTION OF THE STONE IDOL -- DIVIDES THEM ALONG ANCIENT TRIBAL LINES. THE DEBATE IS BOTH HEATED AND CHAOTIC

SUDDENLY, THE IDOL'S RIGHT EAR BREAKS OFF, FALLS ONTO THE DAIS AND SHATTERS INTO TINY FRAGMENTS ...

IN THE ENSUING RIOT, WHICH LASTS ONLY MOMENTS, FOUR DOZEN PIGTS ARE KILLED AND A HUNDRED MORE ARE WOUNDED...

WHAT DID YOU JUST SAY?

SOMETHING HAS GONE TERRIBLY WRONG-- THE SECOND COUSIN OF ALL BATTLES HAS ALREADY TAKEN PLACE AND IT APPEARS THE MEN HAVE COME IN A DISTANT RUNNER-UP-- THERE IS NO SIGN OF THE ONE TRUE POPE AND, JUDGING BY THE LOOKS ON THEIR HOODS, THE OCCUPYING FORCE OF CIRINISTS IS NOT ALTOGETHER LIKELY TO ACCEPT THAT MY REFERENCE TO "STINKING LEZBOS" WAS A WHIMSICAL AND LIGHTHEARTED MALAPROPISM!! NOTHING TO DO NOW BUT...
WHO?
YES
ALL MOTHERS... AND CHILDREN UNDER THE AGE OF FIVE...
THERE'S BEEN AN OUTBREAK OF THE PLAGUE AND WE'RE NOT TAKING...
ANY CHANCES

THERE *IS* NO PLAGUE
HUSH, ALEXANDRA
WHERE WILL YOU TAKE *US?*
TO HEALTH CENTERS NEARBY...
JUST A *PRECAUTION,* REALLY...
YOU'LL BE RETURNED HERE AFTER YOU AND YOUR BABY HAVE BEEN TESTED ...
IT'S BECAUSE WE SAW THE *POPE* ISN'T IT?
ALEXANDRA! HUSH I SAID
WE MUST THINK OF THE *BABY'S NEEDS FIRST*
ISN'T THAT *RIGHT?*
YES MA'AM

MOTHER.
YES, ALEXANDRA.
WHEN HE'S GROWN UP
YOU'LL TELL HIM HE HAD AN OLDER SISTER
... WON'T YOU?

A VISITOR YOU SAY?
I'M
I'M AFRAID I DON'T KNOW HOW THIS... WORKS.
DO YOU CUT MY THROAT OR
OH NO MISS ...
NO NO
A SHARP BLOW IS ALL.
AT THE BASE OF THE SKULL.

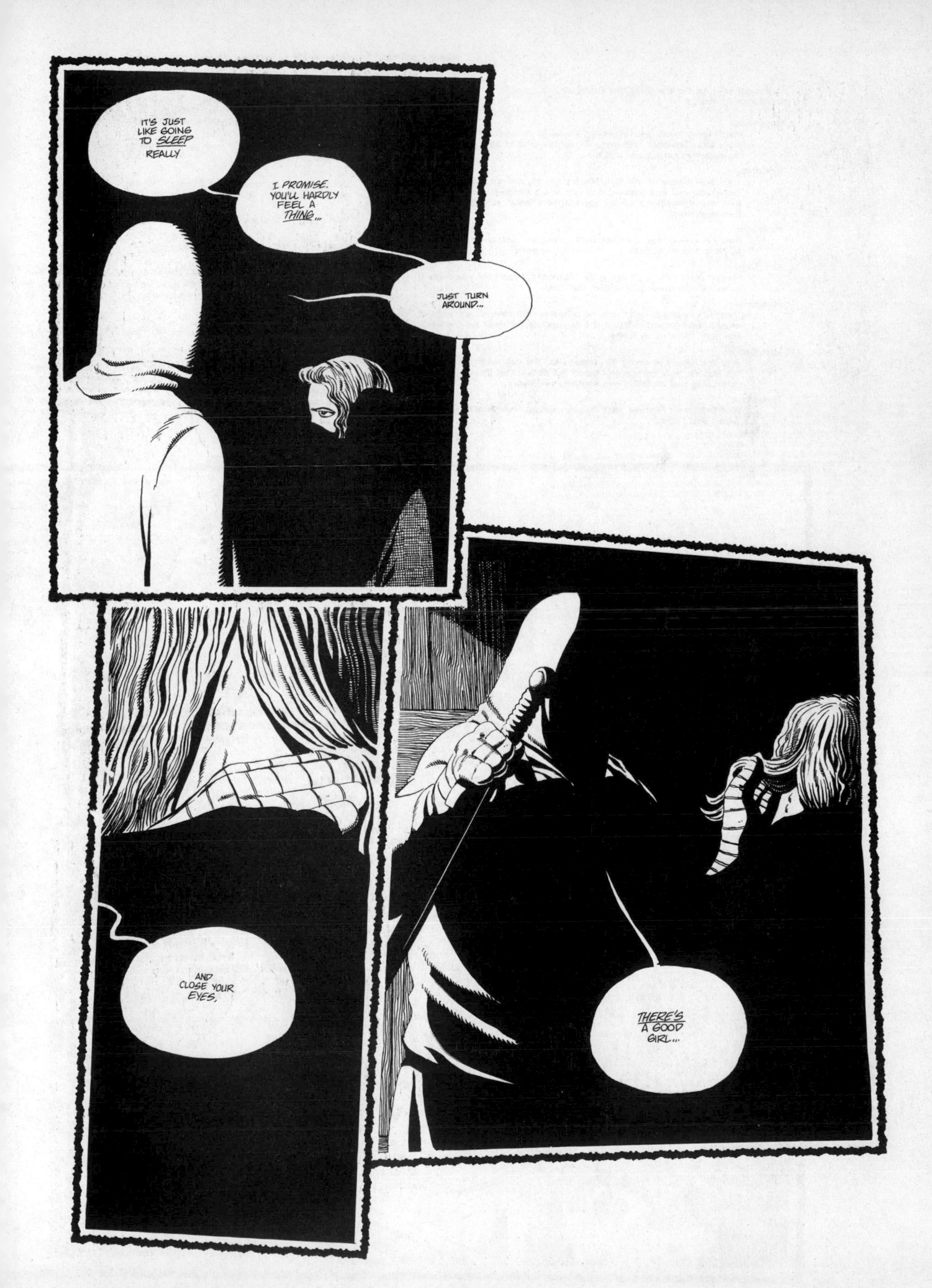
IT'S JUST LIKE GOING TO SLEEP REALLY
I PROMISE. YOU'LL HARDLY FEEL A THING...
JUST TURN AROUND...
AND CLOSE YOUR EYES.
THERE'S A GOOD GIRL...

transcript copy to Her Worship Cirin, Papal Library
transcript continues

Hammond:
don't know what you're asking. I was given specific instructions from Mrs. Thatcher that the gold sphere was to be exactly thirty-three meters across and that the . . .

Mrs. Copps:
I am well aware of the fact, Mr. Hammond, and as I am trying to explain to you, our blessed lady has had a vision that has told her that the dimensions of the gold sphere must now be forty-*six* meters across.

Hammond:
And I'm telling you that that isn't *possible. Look* down there. The forge is half-constructed . . . the mold has been . . .

Mrs. Copps:
Please don't raise your voice, Mr. Hammond, I can hear you perfectly well if you speak in a normal, civil tone.

Hammond:
What I'm saying is that the forge is half-constructed . . . the mold is already half-constructed . . . the dimensions for each are based on the thirty-three meter figure . . .

Mrs. Copps:
And that figure is wrong, Mr. Hammond. The correct figure is forty-six meters. You were instructed at the outset that there was every possibility that modifications might be necessary.

Hammond:
Modifications!! I agreed that if you needed modifications in the *plans,* that I could . . .

Mrs. Copps:
You're raising your voice again, Mr. Hammond. If you insist on shouting, we'll just have to postpone this little discussion until you are willing to discuss things in a calm and reasonable manner.

Hammond:
Yes. Yes. I'm sorry. If I may be permitted to finish my previous thought.

Mrs. Copps:
Certainly, Mr. Hammond. Your opinion in these matters is of the utmost importance. It is, after all, your field of expertise.

Hammond:
Yes. Yes it is. Thank you. You're very kind.

Mrs. Copps:
Much better, Mr. Hammond. Please go on.

Hammond:
Yes. What I'm attempting to point out is that the construction has been under way for several months now, beginning with the excavation for the furnace and proceeding through the construction of the forge, the ladle and the mold.

Mrs. Copps:
Your progress has been remarkable, Mr. Hammond. Our Lady has been very pleased.

Hammond:
Yes. Yes. Thank you. I've had to pioneer several different mathematical disciplines just to initiate the project. Several members of my staff who specialize in architecture, construction and metalurgy have found themselves in the same circumstance. Nothing of this scale has ever been attempted, every element has had to be developed independently . . .

Mrs. Copps:
The Goddess has informed your endeavours at every turn. Her divine inspiration fills each corner of your enterprise.

Hammond:
Yes. I . . . well, whatever. As you can see if you examine the site below us, you can see that each of the three elements, the forge, the ladle and the mold interconnect . . . it is critical that the interconnection be completely seamless if the sphere is to be pure; without blemish or flaw.

Mrs. Copps:
And it is seamless, bless Our Lady, Mr. Hammond. Now it just needs to be made a teensy-bit larger. Forty-six meters to be precise. Not thirty-three.

Hammond:
But it can't be done. It can't be made a centimeter bigger. It can't be made a millimeter bigger. The whole thing is . . .

Mrs. Copps:
Mr. Hammond, you're shouting again.

Hammond:
Yes. I'm sorry. I . . . if you want the sphere to be forty-six meters, then we'll just have to start again. From scratch. We'll have to tear everything out down to the excavation, enlarge the excavation itself . . .

Mrs. Copps:
Goodness, Mr. Hammond, do you really think you have time to do all that?

Hammond:
Time?

Mrs. Copps:
Well, yes. After all, you've promised that the sphere will be delivered by the end of next month. I wouldn't think you'd have time. I think Our Lady's suggestion that you modify what you have done already makes much more sense under the circumstances.

Hammond:
But if you want everything changed. I mean, I can't possibly guarantee a completion date if you . . . it just isn't possible.

Mr. Copps:
Nonsense, Mr. Hammond. You've done splendidly. Now you just have to make a small change in the size. We wouldn't dream of asking you to undo all of your wonderful work. What a waste that would be, Mr. Hammond. Mm? What a dreadful, dreadful waste.

Hammond:
I. I'll see what I can . . . I'll talk to my staff . . .

Mrs. Copps:
That's the spirit, Mr. Hammond. We have the utmost confidence in you.

Hammond:
Thank you.

Mrs. Copps:
Now the second bit of business is your cost over-runs. We simply have to find a way to cut down on all the money you're spending on labour, materials and transportation costs. Goodness, just look at these figures for last month . . .

SCREECH

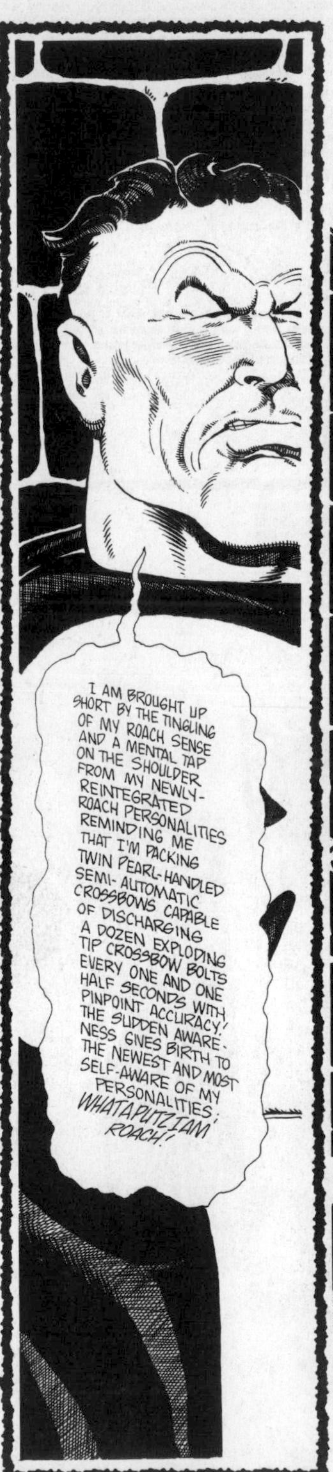
I AM BROUGHT UP SHORT BY THE TINGLING OF MY ROACH SENSE AND A MENTAL TAP ON THE SHOULDER FROM MY NEWLY-REINTEGRATED ROACH PERSONALITIES REMINDING ME THAT I'M PACKING TWIN PEARL-HANDLED SEMI-AUTOMATIC CROSSBOWS CAPABLE OF DISCHARGING A DOZEN EXPLODING TIP CROSSBOW BOLTS EVERY ONE AND ONE HALF SECONDS WITH PINPOINT ACCURACY! THE SUDDEN AWARENESS GIVES BIRTH TO THE NEWEST AND MOST SELF-AWARE OF MY PERSONALITIES: *WHATAPUTZIAM ROACH!*

EAT TASTY SEMI-AUTOMATIC DEATH YOU DILDO JOCKIES
TUNG TUNG TUNG TUNG TUNG TUNG TUNG TUNG TUNG TUNG TUNG TUNG TUNG TUNG
TUNG TUNG TUNG TUNG TUNG
TUNG TUNG TUNG

TUNG TUNG TUNG TUNG TUNG TUNG
TUNG TUNG TUNG TUNG TUNG TUNG
TUNG
SPLAT
TUNG TUNG
TUNG TUNG TUNG TUNG TUNG TUNG TUNG TUNG
TUNG TUNG TUNG TUNG TUNG TUNG TUNG TUNG

GENERAL?

GENERAL WHAT IN HELL IS GOING ON DOWN THERE?
GENERAL?

"PLAGUE"?
AH SAY...
"PLAGUE!"?
HEH HEH HEH
DON'T YOU ALL GO WORRYIN' YO POINTY HOODED HAID ABOUT L'IL OLD ME...
IN CASE YOU ALL HAVEN'T NOTICED AHM THE NEWEST CARD-CARRYIN', DUES PAYIN' MEMBER OF THE L.O.S, HYPHEN F. HYPHEN T. HYPHEN C.
THE LEGION O'SIX-FOOT-TELEPATHIC-COCKROACHES...
US SIX-FOOT TELEPATHIC COCKROACHES ARE AS IMPERVIOUS TO HARM AS AN ORANGE PEEL AT A NAIL-BITER'S CONVENTION

SO IF YOU ALL JUST STEP BACK WHILST I STEP OVER THIS HERE IMPROVISED BARRICADE ...
ON THE OTHER HAND "YOU CAN NEVAH BE TOO CAREFUL ABOUT AN OUTBREAK OF THE PLAGUE..."
"...OR EATING BLUE FRUIT"
AS MY DADDY USED TO SAY
SQUEEE

ASTORIA
YOUR IMPRISONMENT IS NEARLY AT AN END.
A MAN IN BLACK HE'S KILLING US ALL TERRIBLE NEW WEAPONS
WHO IS THIS? WHO DARES?
ONE OF NORMINA ONE OF GENERAL SWARTSKOF'S AIDES
GENERAL SWARTSKOF SHE'S DEAD SHE--
UNGLK!

WHAT IN HELL IS GOING ON DOWN THERE?
YOU ARE UNDOUBTEDLY A MANIFESTATION OF MY DEEPEST INNER FEAR THAT I AM NOT AN AGENT OF THE WILL OF THE LIVING TARIM BUT THAT I AM IN FACT, THE LIVING TARIM HIMSELF AND I HAVE CONDEMNED MYSELF TO OBSERVATION WITHOUT PARTICIPATION IN HUMAN AFFAIRS FOR ALL TIME AS A WAY OF APOLOGIZING FOR HAVING CREATED DEATH
ALSO (UNDOUBTEDLY) I HAVE CREATED YOU AS A KIND OF "EVIL TWIN" WHO IS ABLE TO ACT OUT DEEP-ROOTED URGES SUCH AS TELLING DEATH HE DOESN'T EXIST...
THEREBY CAUSING HIM TO CEASE TO EXIST
PERMITTING ME, THEREFORE TO ELIMINATE DEATH WITH-OUT HAVING TO ACCEPT THE BLAME FOR HAVING DONE SO...

DEATH IS A FACT OF EXISTENCE HAVING NEITHER PHYSICAL NOR SPIRITUAL FORM. YOU ARE (OBVIOUSLY) A METAPHYSICAL INCARNATION OF SOME SELF-DELUDING ENTITY OF ILL-DEFINED PARAMETERS AND MOTIVATION SEEKING TO ESCAPE OR JUSTIFY YOUR OWN SELF-DESTRUCTIVE NATURE BY TAKING THE FORM OF AN IMPARTIAL AND OMNISCIENT UNIVERSAL ARCHETYPE AS A MEANS OF EVADING YOUR OWN ACTIONS AND THEIR CONSEQUENCES ...
ANOTHER EXAMPLE OF THE CIRCUITOUS REASONING BY WHICH YOU, MY EVIL TWIN, PERMIT YOURSELF TO ENCOMPASS ALL THE TRAITS AND QUALITIES OF THE LIVING TARIM -- TO WIT 'AN IMPARTIAL AND OMNISCIENT UNIVERSAL ARCHETYPE' -- WHILE SIMULTANEOUSLY AVOIDING THE CONCLUSION TO WHICH SUCH A DEFINITION MUST LEAD BY ITS VERY NATURE...
LET'S GO THROUGH THAT AGAIN ... SLOWLY!
OF COURSE
I HAVE NOTHING BUT TIME.
WHO ARE YOU?

WHO ARE YOU?
QUIET IN THE CELLS!

IF YOU'RE THE REAL REGENCY ELF...
THEN WHO STAYED IN THE AMBASSADOR SUITE RIGHT BEFORE CEREBUS MOVED IN...
COUNTESS ELAINE SHELLY ...
AND WHAT DID SHE DO EVERY MORNING WHEN SHE GOT OUT OF THE ...
SHE PICKED THE DEAD SKIN OFF OF HER BIG TOE AND SHE ATE IT

BLECCH!

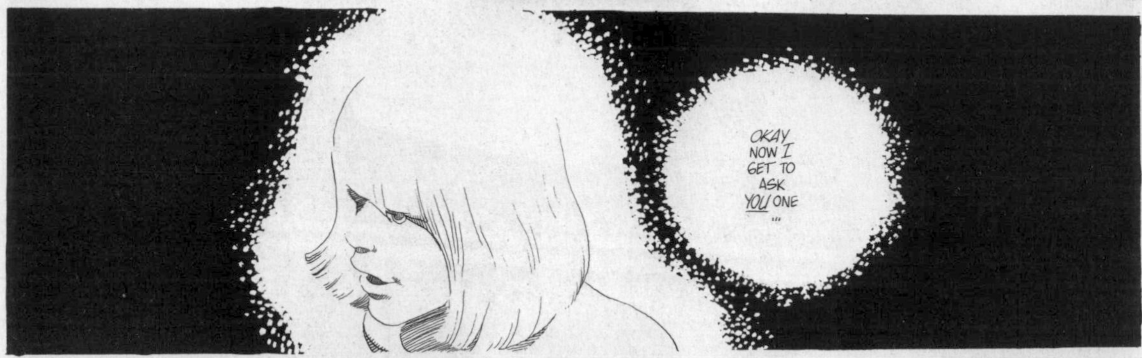
OKAY NOW I GET TO ASK YOU ONE ...

CEREBUS?

WELCOME BACK TO THE SEVENTH SPHERE

SUENTEUS PO...
AT YOUR SERVICE
NEXT: MIND GAME V

OR PERHAPS AT YOUR DISSERVICE ...
IT'S HARD TO TELL.
I DON'T KNOW WHO'S CAUSING ALL THE UNEXPLAINED APPEARANCES AND DISAPPEARANCES
THE MORE UNEXPLAINED OCCURENCES TAKE PLACE, THE MORE PEOPLE BELIEVE IN ILLUSIONISM
THE MORE THEY BELIEVE IN ILLUSIONISM THE MORE EVERYTHING BECOMES AN ILLUSION
MOST ESPECIALLY ME...
AT THE RATE THINGS ARE GOING, IN A FEW DAYS EVERYONE WILL BELIEVE IN ILLUSIONISM

AND I'LL CEASE TO EXIST ...
THE BOTTOM HAS DROPPED OUT OF REALITY ...
I CAN'T EVEN MAINTAIN A PHYSICAL FORM FOR MORE THAN A FEW SECONDS ...
HOW HUMILIATING
I'M JUST FADING AWAY BEFORE MY NON-EXISTANT EYES...
I CAN'T EVEN REMEMBER MY....
MY...?
uh...

SUDDENLY, THE VOICE OF FRET MAC MURY RESOUNDS THROUGH THE ANCIENT CATACOMBS. A HUSH FALLS OVER THE MILLING THRONG, MESMERIZED BY THE CLAN LEADER'S WORDS
"THE GREAT CEREBUS HAS RETURNED," HE INTONES. "HE IS AN IDOL THAT WE WORSHIP.. NOT OF FLESH, BUT OF STONE. NOW THAT HE HAS RE-EMERGED WE MUST RESIST THE URGE TO FOLLOW THE FALSE PATH OF BRAN MAK MUFIN..."
"THE GREAT CEREBUS RETURNS, BIDDING US NOT TO GO FORTH INTO THE WORLD OF THE UNBELIEVERS."

"BUT TO REMAIN IN OUR UNDERGROUND CITIES! HIS RETURN IS A PROMISE TO US ALL."

"SO LONG AS WE REMAIN HERE, HE WILL KEEP ALL THE CLANS SAFE FROM HARM."
"SEARCH YOUR HEARTS" HE CONCLUDES "AND FIND THE TRUTH IN MY WORDS"

THERE IS MURMURRED AGREEMENT AMONG HIS LISTENERS ...

UNTIL A BOLT OF PURPLE LIGHTNING FLASHES FROM THE CEILING, INSTANTLY REDUCING FRET MAC MURY TO A SMALL PILE OF GRAY ASHES ...

WHICH SERVES TO TIP THE BALANCE OF OPINION (SOMEWHAT DECISIVELY) AGAINST THE ADVOCATES OF ISOLATIONISM.

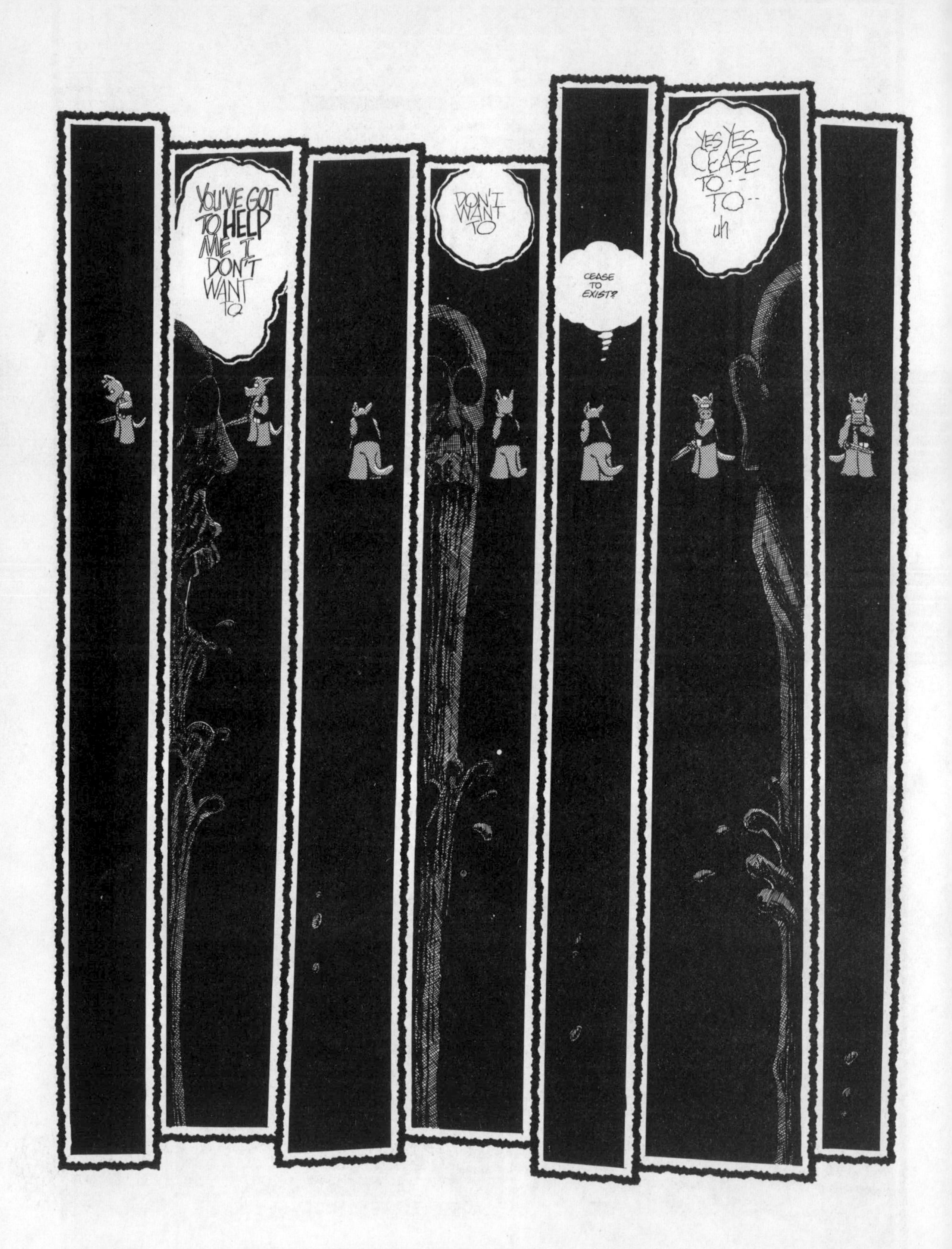
YOU'VE GOT TO HELP ME I DON'T WANT TO.
DON'T WANT TO
CEASE TO EXIST?
YES YES CEASE TO-- TO--
uh

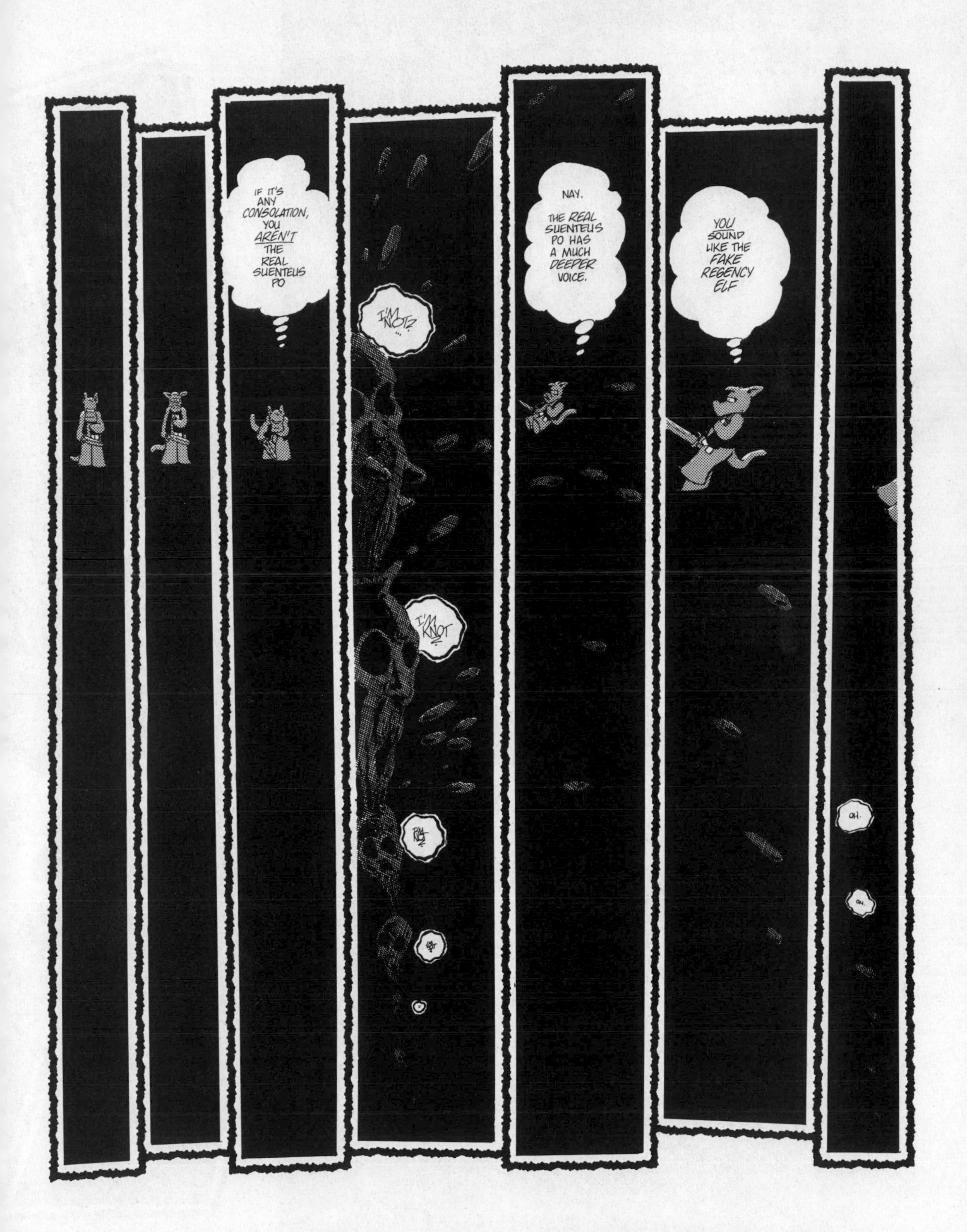

IF IT'S ANY CONSOLATION, YOU AREN'T THE REAL SUENTEUS PO
I'M NOT?...
I'M KNOT
NAY.
THE REAL SUENTEUS PO HAS A MUCH DEEPER VOICE.
YOU SOUND LIKE THE FAKE REGENCY ELF
OH.
OH.

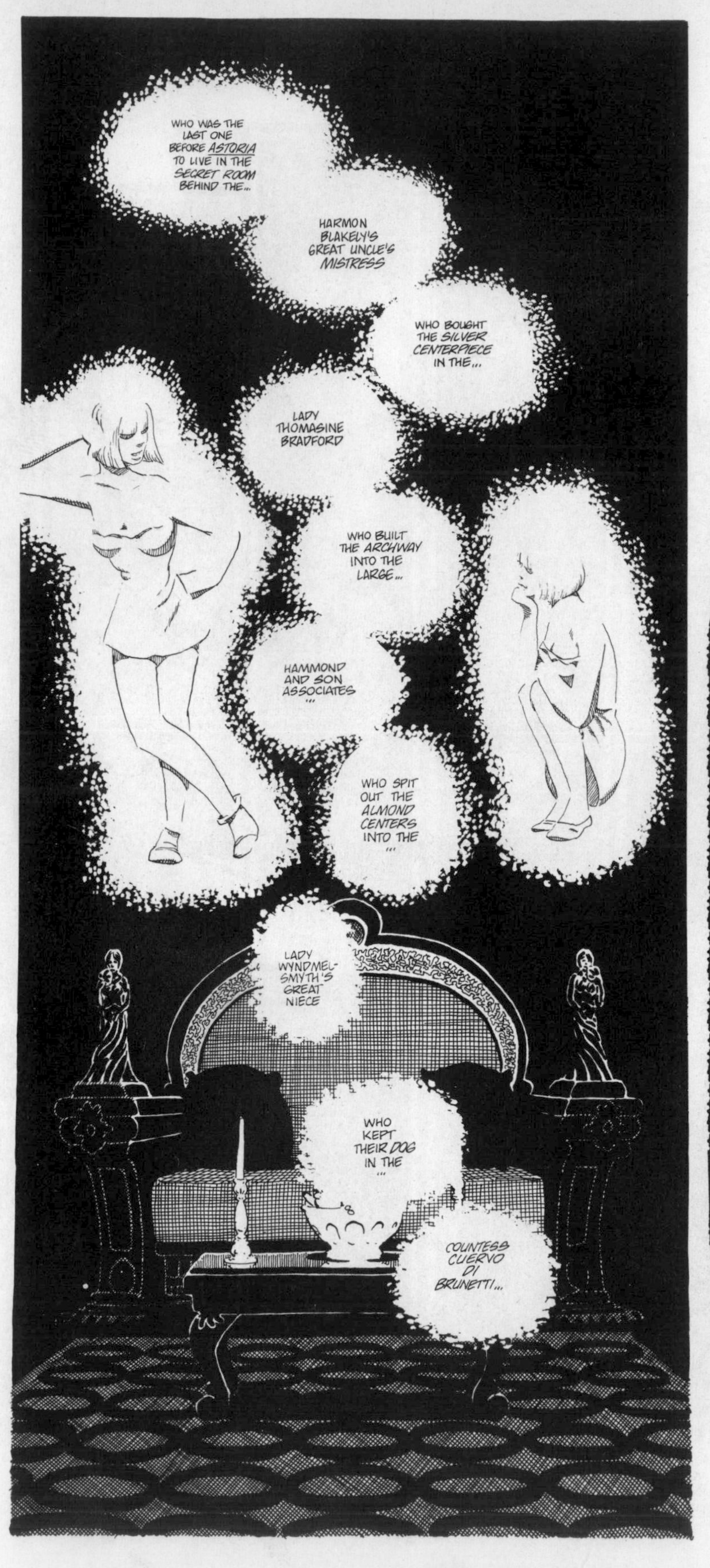
WHO WAS THE LAST ONE BEFORE ASTORIA TO LIVE IN THE SECRET ROOM BEHIND THE...
HARMON BLAKELY'S GREAT UNCLE'S MISTRESS
WHO BOUGHT THE SILVER CENTERPIECE IN THE...
LADY THOMASINE BRADFORD
WHO BUILT THE ARCHWAY INTO THE LARGE...
HAMMOND AND SON ASSOCIATES ...
WHO SPIT OUT THE ALMOND CENTERS INTO THE ...
LADY WYNDMEL-SMYTH'S GREAT NIECE
WHO KEPT THEIR DOG IN THE ...
COUNTESS CUERVO DI BRUNETTI...

VERY GOOD YOUNG CEREBUS
WELL DONE ...

YOU'RE LEARNING FINALLY-- HOW LONG HAS IT BEEN SINCE ANYONE HAS SEEN THROUGH ONE OF MY BETTER ILLUSIONS
TOO LONG I THINK
TOO LONG I THINK
SO-- LET'S ACCELERATE THINGS A LITTLE BIT-- TIME IS OF THE ESSENCE --IF THE BOTTOM HASN'T DROPPED OUT OF REALITY IT IS (AT THE VERY LEAST) SAFE TO SAY THAT THE BOUNDARY BETWEEN TRUTH AND ILLUSION HAS BEEN RUPTURED---

APPEARANCES AND DISAPPEARANCES-- THE TRIUMPH OF ILLUSION OVER TRUTH SO THAT NOW ILLUSION IS TRUTH...
DOUBT-- RADICAL DOUBT FOR THE FIRST TIME ASSAILS THE UBIQUITOUS AND HITHERTO INDOMITABLE MATRIARCHY-- MY AGES-LONG BATTLE WITH CIRIN AT A CONCLUSION
ANCIENT FORCES RACE FORWARD TO ENGAGE AND THEN VANISH-- ANCIENT QUESTIONS ARE ANSWERED-- "WHICH CAME FIRST-- THE CHICKEN OR THE EGG" -- ANSWER? "NO ONE CAME UNTIL THE ROOSTER DID" FASTER AND STILL FASTER AND STILL FASTER
WHAT MANNER OF REALITY REMAINS WHEN EVERYTHING HAS CHANGED? WHEN MY ILLUSIONS CAN BE SEEN THROUGH SO READILY?-- IT'S NOT THE SAME AS YOUR OTHER VISITS HERE-- I AM FILLED THIS TIME WITH THE SENSE THAT MUCH OF THE FUTURE COURSE OF THE SEVENTH SPHERE HINGES ON YOU
YOU'VE SEEN SO MUCH AS KITCHEN STAFF SUPERVISOR, AS PRIME MINISTER, AS POPE OF THE EASTERN....
THIS ISN'T THE SEVENTH SPHERE
AND YOU AREN'T SUENTEUS PO, EITHER
YOU SOUND LIKE THE FAKE REGENCY ELF WITH A HEAD-COLD...

SO WHAT YOU ARE SAYING IS; IF I HAVE ALL THE ABILITIES OF TARIM; IF I HAVE THE NATURE OF TARIM AND ALL ASPECTS OF TARIM, THEN I MUST BE TARIM... AND THAT I AM AVOIDING ACKNOWLEDGING THAT FACT BECAUSE THE IMPLICATION OF DOING SO WOULD REQUIRE ME TO ACKNOWLEDGE THAT THE MESS EVERYTHING IS IN IS A RESULT OF A TOTAL FAILURE OF WILL ON MY PART... THE VERY THING OF WHICH I ACCUSED "DEATH" BEFORE HE CEASED TO EXIST...
EXACTLY. NOT ONLY IS IT YOUR GREATEST FEAR THAT YOU ARE, IN FACT, THE LIVING TARIM (AS YOU'VE JUST ACKNOWLEDGED) ...IT ALSO YOUR GREATEST SECRET HOPE -- THAT THE LIMITATIONS ON YOUR ACTIONS (TO OBSERVATION AND NON-INTERFERENCE) ARE SELF-IMPOSED AND THAT, ULTIMATELY, YOU ARE CAPABLE OF CHANGING THAT WHICH NEEDS CHANGING.

A MOST PERSUASIVE ARGUMENT

IT OCCURS TO ME THAT THE ONLY WAY TO BE CERTAIN OF YOUR THESIS IS FOR ME TO ATTEMPT TO INFLUENCE SOME EVENT TAKING PLACE SOMEWHERE IN THE OMNIVERSE AT THIS VERY MINUTE ...

IF I AM ABLE TO DO SO WITH IMPUNITY, I WILL BE FORCED TO ADMIT YOU'RE RIGHT ... I AM THE LIVING TARIM

ON THE OTHER HAND YOU WERE THE ONE WHO TOLD CEREBUS THAT YOU WERE ROOTING FOR WEISSHAUPT TO ACCOMPLISH THE FINAL ASCENSION

WHICH MILITATES AGAINST ANY CLAIM OF OMNISCIENCE ON YOUR PART AND MAKES THE POSSIBILITY OF YOUR BEING THE LIVING TARIM A REMOTE ONE AT BEST...

WHICH RETURNS US TO THE "KNOW-IT-ALL-WISE-GUY-EVIL-TWIN THEORY"
IF YOU DON'T WANT MY ADVICE PLEASE
JUST SAY SO

SPLUP
SPLUP
SPLUP
SPLUP
SPLUP
PLUP

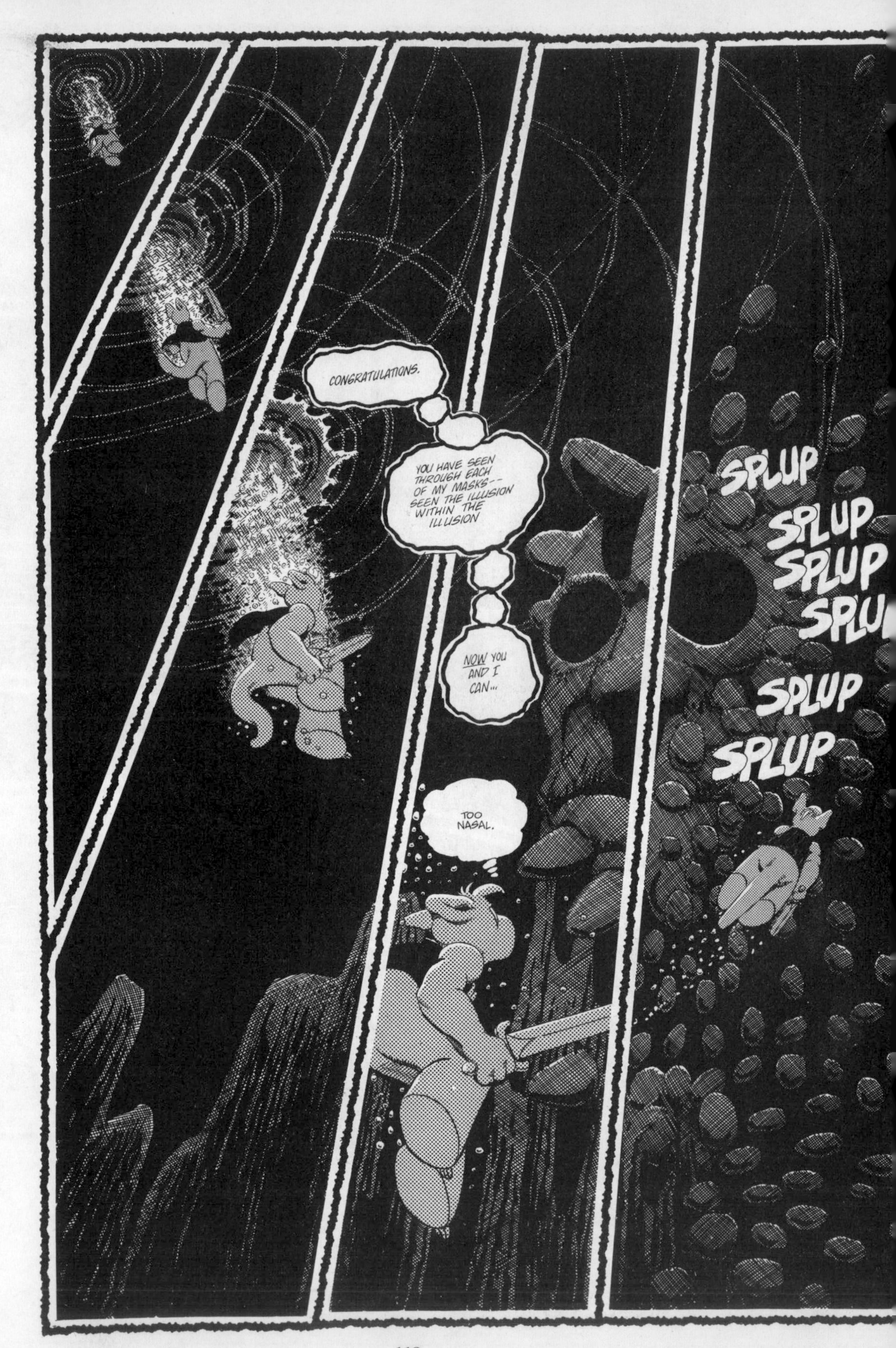
CONGRATULATIONS.
YOU HAVE SEEN THROUGH EACH OF MY MASKS-- SEEN THE ILLUSION WITHIN THE ILLUSION
NOW YOU AND I CAN...
SPLUP
SPLUP
SPLUP
SPLUP
SPLUP
TOO NASAL.

so we got a volunteer for the suicide mission from among the Infertiles, provided her with a crossbow and had her walk straight into the crisis area while we all monitored through her eyes.

Cirin:
And?

General Greer:
It's one man. Dressed completely in black. Design of a skull on his upper torso. The weapons are hand-held. She didn't get close enough for us to get a good look but they seem to be miniature cross-bows, capable of firing dozens of bolts simultaneously. Accurate, too. She was still a couple of hundred meters away and he hit her three times in the left breast.

Cirin:
I see.

General Greer:
She didn't live long enough for us to get a close look at her wounds but from what we could see, they weren't consistent with a usual crossbow bolt wound. They were much larger . . . *much* larger. Somehow the bolt expands or bursts on impact. Anything within a meter of any vital organ is going to be fatal and he doesn't seem to miss by a meter for the most part.

Cirin:
We'll just have to over-whelm him with sheer force of numbers, then.

General Greer:
With all due respect, Great Cirin, there's no evidence that they would get within range no matter how many of them there are. We've already lost all five dozen soldiers we had within the perimeter and we still don't have a clear picture of what we're dealing with. If possible, we'd like to have a few hours to review what we do know and then propose a few different courses of action that we think have a good possibility of success.

Cirin:
Normina . . . General Swartskof.
She's dead then?

General Greer:
I'm afraid so . . . all of our people in that area of the Lower City.

Cirin:
We'll just have to carry on, then.
Nothing must be allowed to threaten the One True Ascension.

General Greer:
Yes, Great Cirin.

Cirin:
Assemble your staff, then, and put your minds to it. I'll expect your recommendations by nightfall.

General Greer:
Yes, Great Cirin. Something I forgot to mention that the Infertile saw on the way in. A number of hand-painted signs were posted around the crisis area.

Cirin:
Signs?

General Greer:
Yes, Great Cirin. They said "Welcome to Roachland." We're not sure what it means.

Cirin:
I see. Thank you, General Greer.
Dismissed.

General Greer:
Yes, Great Cirin.

HAVE A CARE YOUNG CEREBUS...
YOU ENTER THE INNER SEVENTH SPHERE AT YOUR EXTREME PERIL...
FEAR IS A TANGIBLE THING, HERE, WHERE SLAVERING JAWS OF
TOO LOUD.

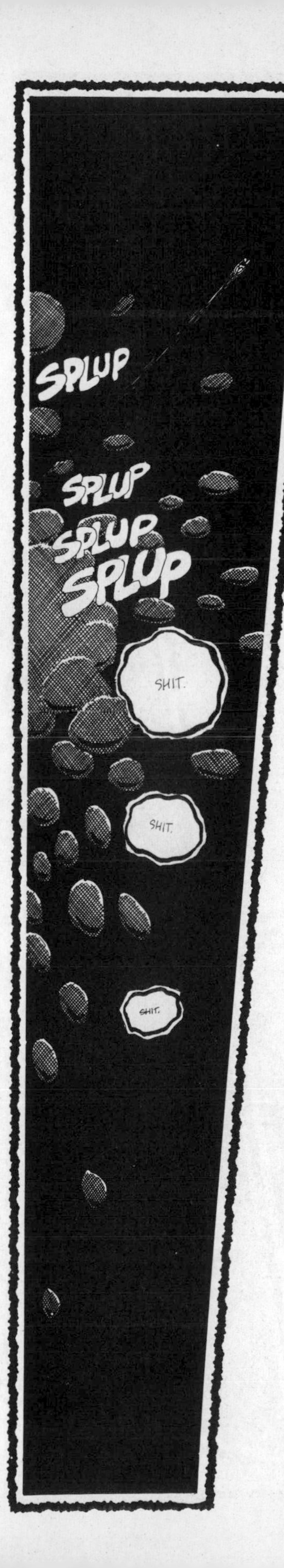
SPLUP
SPLUP
SPLUP
SPLUP
SHIT.
SHIT.
SHIT.

YES, O GODDESS OF LIFE ETERNAL ...

I HEAR.
AND I OBEY.

VERY WELL.
YOU HAVE GAINED THE INNER REALMS OF THE SEVENTH SPHERE ...
MY BEST ILLUSIONS ARE WASTED ON YOU
JUST ANSWER ONE QUESTION FOR ME...
WHERE DO YOU THINK YOU'RE GOING?
UP.
UP?
THAT'S IT? "UP"?
AYE.
CEREBUS ALWAYS PROMISED HIMSELF THAT IF HE MADE IT BACK TO THE SEVENTH SPHERE ...

CEREBUS ARE YOU LISTENING?
CEREBUS?!
NOBODY LIKES A WISE-GUY AARDVARK!!
THE MOON? TWO HUNDRED AND THIRTY THOUSAND MILES ...
YOU MIGHT AS WELL ASK "HOW HIGH THE MOON?"
... THAT CEREBUS WOULD FIND OUT HOW HIGH UP IT GOES ...
HOW HIGH UP IT GOES?!
THE SEVENTH SPHERE?!?
HAH!

THAT CAWN'T BE RIGHT.

YOU SEE? THAT'S WHAT I THOUGHT... BUT MRS. COPPS IS MOST INSISTENT -- IF I BEGIN TO DISMANTLE THE SPHERE SHE SAYS THERE ISN'T TIME...

AND IF I TRY TO PROCEED WITH CONSTRUCTION SHE SAYS IT'S THE WRONG SIZE

SO WHAT ARE YOU DOING?

HALF OF MY BLOODY STAFF IS TEARING THE DAMN THING DOWN!
AND HALF OF THEM ARE BUILDING IT BACK UP!!

I'M SORRY I DIDN'T MEAN TO RAISE MY VOICE
PERFECTLY UNDERSTANDABLE MR. HAMMOND
PLEASE RETURN TO YOUR DUTIES I'LL SEE WHAT CAN BE DONE ...

OKAY! OKAY! IT'S ME -- THE REAL SUENTEUS PO!
RIGHT?
RECOGNIZE THE VOICE?
DEEP?
NOT NASAL?
NOT LIKE THE FAKE REGENCY ELF WITH A HEAD COLD?
NOT TOO LOUD?
CAN WE TALK ABOUT THIS?
I MEAN WOULD YOU MIND JUST HOLDING ON A SECOND WHILE WE TALK THIS OVER? ...
NAY
CEREBUS IS GOING UP.
UP! -- WHAT'S SO GREAT ABOUT UP, ANYWAY?
WHY DOES EVERYONE TODAY THINK SIDEWAYS IS OVER-RATED ?
LOOK! I CAN SAVE YOU A TRIP! ...
I HAVE POST CARDS
CEREBUS WANTS TO SEE FOR HIMSELF

NEXT; MIND GAME VI

Congratulations.

Through your single-mindedness of purpose, you have achieved your goal and have now arrived at "up". That is to say, you have arrived in the Eighth Sphere which, technically speaking, is as "up" as one can* get *in the Seventh Sphere.

Aye. Cerebus has been here before.

If you are referring to our curious little exchange at the time of your "kidnapping", no. In point of actual fact, that was* not *the Eighth Sphere. Rather, it was something of a "suburb" of the Seventh Sphere. Without going into great and extraneous detail, it is as I described it at the time; an environment supplied by individual imagination. A series of portents and omens (most of them thoroughly discreditable) had suggested the possibility (however remote) of your emergence as a "key player" in the hurly-burly hustle and bustle of our many-hued multiverse. Quite frankly, I was curious to see what manner of environment your Essential Self might fashion if given the opportunity. If you will recall, you contrived a sort of endless gray slide for yourself, both incomplete and transitory, at the conclusion of which, you vanished without a trace into a small patch of gray bubbles, leaving no discernible change through your passing, suffering neither real nor ostensible ill-effects in the process. I will admit that this made your tenure as Prime Minister a more than slightly predictible anti-climax, but then a satisfied curiosity is the poorest bedfellow of an exquisite anticipation.

Huh?

In the Eighth Sphere we don't provide translations for the slow-witted, I'm afraid. "Up" comes without attendant footnotes. Grasp what you can and try not to breathe through your mouth so much. Apropos nothing; might one hazard a guess as to the reason for your reclined head, pointed toes and arched back? Dare one suppose that you are still in pursuit of "up"?

Aye.

Well, you can spare yourself the potential dislocation of important muscle groups. It is a characteristic of the Eighth Sphere that there is no "down" or "up". There is only eternal darkness in all directions, without interruption or signpost. If you will attempt to look "down" you will notice that there is no vital impression remaining of the Seventh Sphere to mark, for you, even a delusion of progress. This, to put it as succinctly as possible, and at the risk of being redundant, is as "up" as you get.

*Okay, so what's the Eighth Sphere **for**, then?*

For? An interesting — albeit moronic — question. You might as readily ask what *you* ***are "for". Or against if it comes to that. Each person's Eighth Sphere is his own, to have and to hold, the repository of all he has been, all he is and all he will be. You can relive any point of your life. A second can be experienced for a decade. Or you can skim the sum of your cradle-to-grave existence in the blinking of an eye. You can pursue a single emotion; witness the nativity of one of your beliefs, watch it take root, listen to it evolve over years, betrayed, transmogrified and indicted; defended, revived and reprieved in your own words. You can witness its expiration or see it outlive you. For the literal-minded there is no proof, save by example, so let's start with an easy one until you get, as they say, the "hang" of it.***

"Joy"

Here's one your earliest experiences with the emotion of Joy. You're a mere lad of eight summers and you're on your way home after an industrious afternoon of stealing baked goods found cooling in open windows. You have just finished consuming a delectable cherry pie, its filling enhanced with the unexpected tartness of lemon juice. You are passing through a grove of lilac trees. It is late spring and a warm wind suddenly springs up from the south, shaking the blossoms of the surrounding trees with a rush of sound like a wave on the ocean, so that the cream-white and tiny petals fill the air around you like some brilliant, magical snow-fall as far as the eye can see. There is no one threatening you, no one taunting you about your appearance or your clothing and, for the first time, a warmth and an inner peace descends over you. You are relaxed and happy. You stand there, entranced and inwardly pleased, for several uninterrupted minutes.

Moving ahead a few years, we come to the (ahem) be-heading of your first Borealan. Not a cause for Joy among even the least respectable persons of my acquaintance, but, then, the Eighth Sphere is a mere catalogue of impressions and memories and nothing more; passing no judgement; offering self and self alone for each individual's examination. You had practised with your sword in secret for weeks against the day when you would challenge this hirsute and haberdasherial monstrosity to a, as the cliche-infested might have it, battle to the death. As your sword cut smoothly through artery and bone; as a spray of hot blood splattered against your face

(Good Lord I Think I'm Going To Be Sick)

and a triumphant cry arose from the surrounding spectators, your mercenary "chums" (indistinguishable as they might be from your opponent to the careless and civilised eye), a great weight of fear lifts from your shoulders like a bird taking wing and your spirit soars within you.

A few years further on you have deserted your current military commitment and have trudged through trackless wilderness for many days, hunting small game for your supper, sleeping in the hollows of trees, in constant terror that you are being tracked and will be made to suffer the consequence of your desertion. The purpose of your journey is to stand upon the Wall of Tsi and gaze into the lands of the T'Capmin Kingdoms. It had been one of your father's favourite expressions whenever you had expressed an inkling of ambition or evinced an interest in any sort of achievement; "Sure and then you'll stand on the Wall of Tsi." The risk you had undertaken was considerable and the impediments you faced virtually insurmountable. Now, many days later, without a coin in your pocket or notion of what the next day might bring, you breathe deep the biting north wind, redolant of evergreen and wood-smoke and for the first time you feel a sense of total personal achievement.

And then, the most Joyous moment in your principally joyless existence . . .

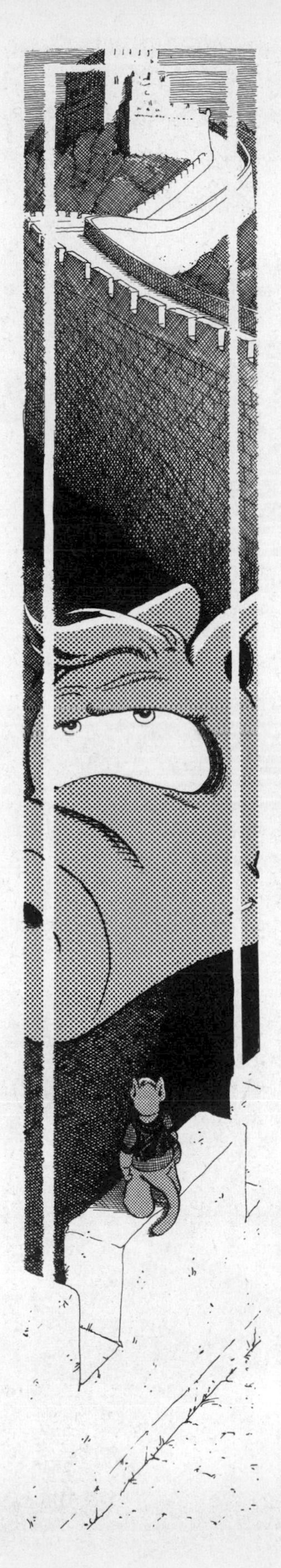

I WILL TAKE YOUR ABRUPT DISAPPEARANCE AS A VOTE OF CONFIDENCE IN MY "KNOW-IT-ALL WISE-GUY-EVIL-TWIN" THEORY...

I KNOW EXACTLY WHAT YOU'RE TRYING TO DO MRS. THATCHER
DON'T THINK FOR ONE MINUTE THAT I DON'T
MRS. COPPS?

BENEATH THE RED MARCHES, THE COMPLETE CHANGE IN PREVAILING OPINION IS VIRTUALLY UNIVERSAL AND INSTANTANEOUS.... THE CLAN ELDERS AND THE CREAM OF PIGTISH MANHOOD TURN-- WORDLESSLY-- FROM THE SCENE OF FRET MAC MURY'S DEMISE, AND BEGIN STRIDING PURPOSEFULLY, AND AS ONE, IN THE DIRECTION OF THE LONG-SEALED ARMOURY...

THE SEMI-AUTOMATIC CROSS-BOW FACTORY WILL GO IN THAT BUILDING OVER YONDER...
RIGHT NEXT TO THE CHURCH OF THE LIVING CEREBUS

The first night you spent with Jaka; when she had agreed without hesitation to accompany you; to be your life's companion. You had dozed briefly, drained by the excitement of that most enchanting day's events. And then you awoke to find her looking at you. Without a word, she had taken you into her arms. There was the light, sweet scent of perfumed soap on her skin and in her hair; the reassuring pressure of her arms enfolding you; and the delicate warmth of her breath on your eyelids.

For the first time in your conscious memory; for the first time in fact, since you were a baby; a single tear, full and warm, rolled down your right cheek and you fell into a very deep and entirely dreamless slumber . . .

YOU'RE TRYING TO TAKE OVER THE ONE TRUE ASCENSION!

YOOOO HOOOO
OTHER EEELLLF
COME OUT COME OUT WHEREVER YOU ARE.

You can feel the drift, can't you? The momentum and the tide of tracing a single emotion through the days of your life. It is a rare gift and a momentous achievement, The Eighth Sphere. And you are not limited to those experiences merely past or simply present. Your future *resides here as well. How wonderful a discovery for someone like yourself, don't you think? You seek "up", but "up" (all things considered) is really a linear projection; a solitary and finite thread. How much richer a tapestry might serve you when "up" is woven in intricate patterns of "forward", of "beyond" — when your progress is not limited to displacement or relocation; when transcendence itself is within your grasp. Experience now, the greatest and most enduring Joy of your adult life a few years hence.*

A word of caution. Remember, what I said about a satisfied curiosity for it is nowhere truer than in the exploration of one's future. Days yet to be should be sampled and savoured in trace amounts. The demarcation between curiosity-seeker and lotus-eater is paper-thin.

Regard.

CLAP CLAP CLAP
CONGRATULATIONS MR. PRIME MINISTER

CLAP CLAP CLAP
BRAVO!
BRAVO!!

CLAP CLAP CLAP
YOU'VE MADE US ALL VERY PROUD TO BE IESTANS TODAY, SIR!

CLAP CLAP CLAP

Gentlemen of the Legislature
Today is a great day in the history of our beloved city-state! For the first time in nearly three years we are free of the oppressive occupation of the hated Cirniots
(applause)
Our armed forces have acquited themselves, Ixotan and

You're not supposed to react with your conscious mind in the Eighth Sphere. It creates alternate time-lines, parallel realities and other manifestations of sloppy psychic house-keeping.

It isn't Cerebus' future. Cerebus dies alone, unmourned and unloved.

Ah. The Judge.

Aye.

In a way, I'm quite surprised that you would take him at his word, considering that he isn't Tarim and that the Moon hardly qualifies, even among the most liberal-minded, as Vanaheim. In another way, I'm not surprised at all; you're the sort of fellow who would believe a street-corner horoscope before he would believe his own heart.

There's no ceiling on the Legislature.

For you? No. Are you sure you don't want to stick around and read the speech?

Nay.

You get a twenty minute standing ovation at the end.

Nay.

Well, what about Blakely's speech where he hails you as the greatest warrior/statesman Iest has ever seen, apart from (ahem) Suenteus Po himself.

Nay.

No. Don't tell me. Let me guess.

"Up"?

Aye.

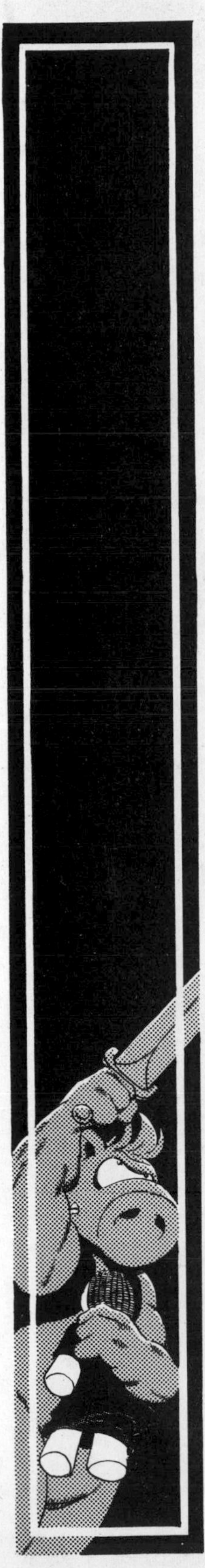

Everyone tries to distract Cerebus. If it isn't you, it's Lord Julius and if it isn't Lord Julius, it's Astoria. Or Elrod. Or someone who wants to paint a picture of Cerebus. Any time Cerebus decides to do something, someone comes along and gets Cerebus to do something else. The more people Cerebus is in charge of, the more distractions there are and the less Cerebus does what Cerebus wants to do and the more Cerebus does what someone else wants Cerebus to do. Well, Cerebus isn't in charge of anyone any more, so Cerebus is going to do what Cerebus set out to do. Cerebus is going up and Cerebus is going to find the top of the Eighth Sphere or the Ninth Sphere or the Tenth Sphere or whatever and wherever it is. Up is up, even if everything else is black. It's like climbing a mountain. You know you're at the top when you run out of mountain and until you run out of mountain, you keep climbing.

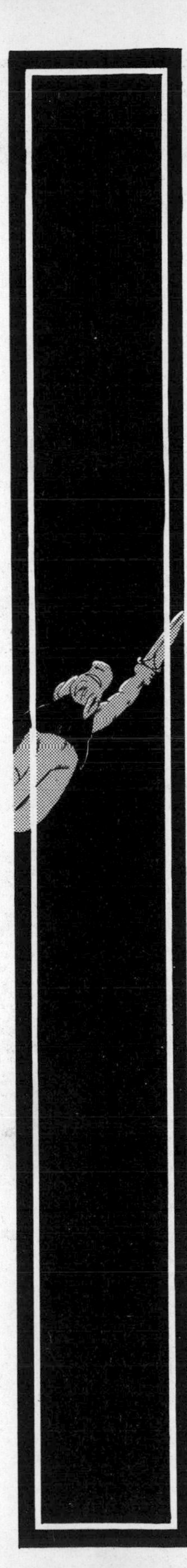

Well said, Well said, indeed. For a fellow whose progress to date might most flatteringly be described as "meager" this smacks of insight writ large. The most distinctive trait of the fool is his submission to distraction and diversion and an abiding faith that they are inescapable matters of course. His quite tedious refrain; "life is what happens when you're making other plans." Conversely, it is a cornerstone of wisdom that obstacles exist to be bypassed; and where that isn't possible they are to be overcome or eliminated entirely. "The more worthwhile the Road, the more seductive will be those paths divergent from it."

"Up", then, Young Cerebus. The final hurdle of your Eighth Sphere has been surmounted and a new reality beckons.

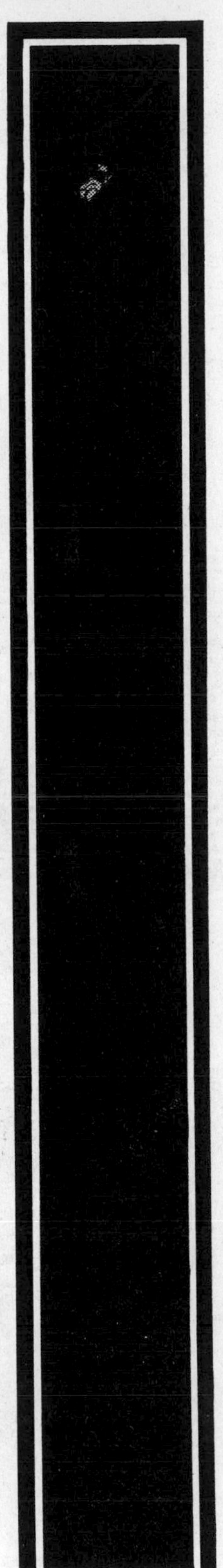

That new reality is a story. Or, rather, THE story. The story that was ancient when the multiverse itself was young. Parts of the story will seem as familiar to you as your own name. It begins (as do all things) with awareness. What You Seek is above you; it is indistinct and shapeless, but also irresistible. Your progress toward it is sure and steady; your urge to proximity an article of faith and your belief in the accessibility of What You Seek unshakeable and profound. You are of it; and it is of you. As you rise toward it, faster and still faster, its shape resolves into a tiny pin-point of light, surrounded by a great and terrible "otherness" for wont of a better definition. Your Curiosity focuses into Desire and Desire into a nameless Hunger for What You Seek.

As you draw closer to it, a new perception of its attraction occurs to you. You are moving toward it at a quite unnatural speed and you find it almost beyond your power to slow your forward momentum. Resisting that pull with every fibre of your being, you retreat until the surrounding darkness is all that there is and What You Seek is once more indistinct and without form.

And there you consider the problem at hand for a very long time, indeed.

At last you move forward again. Slowly.Cautiously. Resistant to the other-worldly attraction which attempts to draw you to it, like iron-filing to magnet. In this new reality, this story, you have a far greater control over your physical presence than you had previously. Consciously, you begin to loosen the bindings between those particles and fragments of which you are constituted. You double, then triple, then quadruple your own mass, reducing your density, pushing relentlessly, carefully, forward; your muscles taut and straining to maintain control. Size is the game here. That Which You Seek is large and it is that size which exerts to much force on you. The pin-point of light enlarges and becomes a flattened disc. Shimmering halos of energy become visible around it. You make yourself larger still.

The force exerted on you is nearly unimaginable; IS unimaginable; increasing exponentially with each incremental move forward. Unexpectedly, a vision swims before your eyes. The disc is a disc no longer, but a platform of some kind, scarcely visible, viewed from its underside. Circling it is a ring of gray and insubstantial matter; tiny gaseous globules in orbit around it. What You Seek is warning you. The communication is without malice and is not a threat; rather it is a statement of fact. If you venture any closer, you will be destroyed. What You Seek is larger than anything you could imagine. Its power a terrible thing scarcely contained within it. Increasing your mass, you retreat to a safe distance and ponder this new piece of intelligence.

The problem is maddening. The thing you seek is so close, you feel you could reach out and touch it. You feel it is your immutable destiny to do so. You have not come this far and at such a cost merely to turn around and go back. There is a solution. Of this you are certain. Now, no longer a game of size, a game of mass, a game of density, it has become, instead a contest of wills. You focus on That Which You Seek as if your gaze alone might bring it closer or narrow the distance between you. Just as it feels as if your mind itself will explode from the strain . . .

A small duplicate of yourself forms above you, perfect in every detail. In that moment of genesis, your awareness divides. You are Cerebus and you are the small duplicate of Cerebus. You are Cerebus and you are the unimaginably large version of yourself you can sense beneath you, counter-balancing the gravitational pull of That Which You Seek. The balance of size and density and mass are shifted and redistributed by you; by your smaller and larger manifestations. You begin to rise like an air-bubble in water, able to control your progress even as you submit to the attraction of That Which You Seek. The disc grows larger. And that's the end of the story. The original *story, anyway, where you reach a point within thirty-six million miles of That Which You Seek and then are frozen into a stasis once more, a mere speck scarcely visible against the disc which virtually blots out the darkness around you.*

In this story, you rise to the edge of the platform and then above it. You can let your arms go limp. Missy and your sword are safe in the custody of your other self far below. The sculptured figures on the platform are familiar to you, of course. And now, it's time we meet face-to-face. You, Cerebus . . . and I, Suenteus Po . . .

WELCOME.
LET'S PLAY SOME CHESS, SHALL WE?

SOMEWHERE VERY HIGH UP AND VERY FAR AWAY, THE DEADLY CHESS GAME BEGINS AS THE LIVING CEREBUS BATTLES FOR CONTROL OF THE MANY-HUED MULTIVERSE...
CIRIN'S ASCENSION IS JEOPARDIZED BY INTERNAL CONFLICTS WITHIN HER OCCUPATIONAL GOVERNMENT
TWO BLOCKS AWAY TO MY LEFT A CITIZEN OF ROACHLAND THINKS OF ASSASSINATING ME AND THROWING HIMSELF ON THE MERCY OF THE CIRINISTS ...
HE TURNS TO MENTION IT TO HIS COMPANION ...
TUNG
SMUT
HALF A SECOND SLOWER ...
AND I'D'VE HAD A CONSPIRACY ON MY HANDS...

THERE ARE *THREE* AARDVARKS ...
YOU.
MYSELF.
AND *CIRIN*
KING'S PAWN TO KING FOUR.

It has been my policy not to interfere with your life. Capricious aspects of my consciousness have, on occasion, violated that rule. But it is only now, now that you have seen the futility of grasping after power and wealth and notoriety that it seems worthwhile for me to interact with you. The greatest danger is that our discourse might alter the future course of events and your place in them. But, frankly, in recent days, very little has been unfolding in the manner proscribed. As a precaution, I will limit the information which I impart to you to self-description. This minimizes the risk of some catastrophe taking place, but it in no way eliminates that chance. You have a reform-minded spirit. In many of my lives, I have been a reformer and if you listen to me carefully and consider, thoughtfully, what I say, there is every possibility that you might avoid the mistakes and pitfalls I have faced. There is nothing wrong with making mistakes, but one should always make new *ones.* Repeating *mistakes is a hallmark of dim consciousness.*

Reform is the centerpiece of the story I told you. Formation and reformation. The fragment detaching from the whole; its disposition and course. All that encompasses human experience is contained therein. Consider the pawn. It has detached itself from the larger entity; disengaged itself from its King and embarks upon a course of action, a series of events and interchanges and discourses. Chess, you see, was never invented. Chess is merely The Story seeking a physical manifestation and finding it. For as long as there has been Awareness, there has been Chess. Each entity maneouvers its pieces on many boards simultaneously and each entity is itself a chess piece of many aspects. We choose our conduct in the many games in progress. In some, we are neutralized; and are impediments only. In some, we are more active, influencing and shaping the nature of the game until it becomes our own. In some, we are mere casualties; early victims of the ebb and flow of events, watching that which unfolds from the sidelines; powerless to intervene.

As the configuration of the board changes, so does the existence and disposition of the pieces.

My first life as a Reformer, the Judge told you about.

BEFORE THE SEALED ENTRANCE TO THE ANCIENT ARMOURY, FIGHTING BREAKS OUT ONCE MORE.
THOSE SUDDENLY DOUBTFUL OF THIS COURSE OF ACTION TURN ON THEIR COMRADES-- CONCEALED DAGGERS FLASH IN THE DIM TORCHLIGHT...
BLOOD MINGLES WITH SWEAT-- THE CRUSH OF HUMAN FORMS MAKES A LIVING HELL OF THE NARROW CORRIDOR... IT IS UNCLEAR WHO IS DEFENDING THE SEALED DOORWAYS AND WHO IS ATTEMPTING TO DISLODGE THOSE DEFENDERS ...
THOSE WHO LOSE THEIR BALANCE ARE SOON CRUSHED UNDERFOOT-- THEIR SCREAMS AND GROANS ECHOING AND RE-ECHOING IN THE EARS OF THE LIVING...
THE FRENZY OF BLOOD-LETTING RISES TO A MANIACAL FEVER PITCH...
LIKE MOTHS TO A FLAME, THE LATE ARRIVALS PRESS FORWARD.

He is well-named for he seems unable to relate a story without passing his judgement on it. Although he purports to be impartial and dispassionate, he reveals himself through his interpretation of the events related. He passed over the Golden Age of the Sepran Empire, dismissing it with a few phrases about being a loosely-knit collection of hamlets, ports and provinces. It was, I assure you, quite nearly Vanaheim on Earth. The greatest freedom for the greatest number, government control so decentralized as to be almost non-existent. I was a Noble, living in a disputed area near the Red Marches at the time that taxation was introduced; tribute paid directly to a governing Church body in Serrea. Although I was dismissed as an hysteric, I knew the effect this would have on people's personal freedoms. Unable to convince any influential personage of the dire future in store, I resolved to raise my own army and conquer the city-state of Iest and institute the same freedoms and policies which had allowed the Sepran Empire to flourish for many, many years. His description of the raising and training of my army is accurate in every detail. He failed to mention that I had taken the name Suenteus Po in tribute to a great historian who had led the Sepran Empire away from the pitfall of centralized government several hundred years before.

THE ANCIENT SEALS BREAK OF THEIR OWN ACCORD WITH A DEAFENING AND EXPLOSIVE SOUND...
FASTER THAN THE EYE CAN SEE, THE IRON AND STONE DOORS, EACH WEIGHING SEVERAL TONS, SWING OUTWARD.
THOSE IN THEIR PATH ARE FLATTENED AGAINST THE RETAINING WALLS ON EITHER SIDE, BEFORE THEY CAN UTTER A SOUND, CRUSHED INSTANTLY BEYOND RECOGNITION.
AS THE ECHO OF THE CONCUSSION ROLLS THROUGH THE UNDERGROUND CHAMBERS SILENCE DESCENDS OVER THE SURVIVORS.
WITHOUT DEBATE, THEY ENTER THE ANTIQUITOUS STORE-ROOM...
... STEPPING CAREFULLY OVER THE RIVULETS AND POOLS OF BLOOD ...

He said that when my army was prepared, I raised my hand, pointed east and said one word; "Kill"

Again, he reveals himself, for in point of fact, the word that I invoked was "Freedom."

Consider the difference and the implications of that difference. For it is the nature of the judgemental to see only Death in Freedom and to see them as interchangeable. Freedom to them is latitude, self-indulgence, mindless willfulness and the imposition of the chaotic on their precious sense of order. He saw me as saying "You are free to kill and cause wide-spread destruction. Do as thou wilt." What I was saying was "What we have in this place is dying, atrophying, becoming corrupt. We must go elsewhere and begin anew."

A CITIZENESS OF ROACHLAND PEERING AT ME THROUGH A RAIL FENCE, THINKING OF PERFORMING UNSPEAKABLE ACTS OF SEXUAL PERVERSION UPON MY ROACH-PERSON FOR DAYS AND DAYS UNTIL, IN A MOMENT OF WEAKNESS, I DESERT THE LIVING CEREBUS AND MAKE HER THE QUEEN OF ROACH-LAND ...

TUNG
SMUT

THE CLOSEST CALL YET...

SHE WHAT?!!

SHE... SAID THAT YOU HAD HAD A VISION --THAT THE SPHERE SHOULD BE FORTY-SIX METERS ACROSS INSTEAD OF THIRTY-THREE...
SHE INSTRUCTED MR. HAMMOND TO MAKE THE NECESSARY...

THAT IDIOT!!
THAT-- THAT-- CUNT!!

IT WAS A DREAM I TOLD HER ABOUT

I WAS JUST MAKING CONVERSATION

MM.
I RAWTHER SUSPECTED THAT WAS...

I WANT HER HEAD.
SHAVED.
AND THEN I WANT A SIGN READING
PENIS-BRAIN.
HUNG.
AROUND HER NECK.
AND THEN I WANT HER FROG-MARCHED
TO THE FILTHIEST.
UGLIEST.
MARKET PLACE IN THE LOWER CITY.
I WANT HER STRIPPED.
NAKED.
AND THEN FLOGGED TO WITHIN
AN INCH
OF HER MISERABLE.
CUNT.
LIFE.
YES.
GREAT CIRIN...

IT WAS AN ILLUSION I TELL YOU
CEREBUS IS DEAD

NO. HE'S ALIVE...
HE'S THE LIVING TARIM
HE HAS TO BE.

BLASPHEMY.

IF IT IS HIM WHERE DID HE GO? AYE? ANSWER ME THAT. WHERE DID HE GO?

HE'S PLAYING CAT AND MOUSE WITH THEM-- THAT'S ALL
THE MINUTE THEY LET THEIR GUARD DOWN ...

I'M TELLING YOU HE NEVER EXISTED IN THE FIRST PLACE
IT'S ALL A TRICK.
AN ILLUSION

SURE... IT'S THE CHURCH-- WEEDING OUT THE UNBELIEVERS ...
THAT'S ALL...

WHAT IF HE IS TARIM? WHAT IF HE'S TESTING US. TESTING OUR FAITH?

OH, HE'S THE LIVING TARIM, ALL RIGHT
HE'S FULFILLED THE PROPHECIES ...

OF COURSE OUR FAMILY HAS ALWAYS BEEN PIVOTAL IN THE BUILDING TRADE IN IEST, SO I'M CERTAIN CIRIN WILL BE...
MRS. COPPS?

THATCHER.
WHAT IS IT? ... I'M VERY BUSY.

THE DIRECTOR OF INTERNAL HARMONY WOULD LIKE A WORD WITH YOU.

THE DIRECTOR OF INTERNAL HARMONY? WHAT THE HELL DOES SHE WANT?

I'M NOT COMPLETELY CERTAIN...

BUT I BELIEVE IT INVOLVES A SHORT VISIT TO THE LOWER CITY...

WHAT ABOUT LAYING WASTE TO THE TEMPLES, AYE?
WHAT ABOUT THAT?

THAT COMES LATER--
AFTER THE FIRE AND FAMINE AND WHAT-NOT

WHAT IF HE NEVER COMES BACK? THIS... COCKROACH MAN...

PUNISHEROACH.

LISTEN-- WHO ELSE CAN WE BELIEVE IN? WE HAVE NO CHOICE...

WHEN THE REAL TARIM COMES BACK, HE'LL TELL YOU YOU ALWAYS HAD A CHOICE
AND YOU CHOSE WRONG...

WE'LL GO ALONG WITH HIM, IS ALL-- FOR THE TIME BEING ...

I THINK TARIM-- THE REAL TARIM-- WANTS US TO KILL THE PUNISHEROACH...
IT'S A TEST ...

WE'RE ALL QUITE MAD...
AREN'T WE?

It is as if the Goddess herself has unleashed the forces of Chaos upon me. I am betrayed on all sides by those I have trusted implicitly. I feel the over-whelming need once more, to have Astoria brought to me; to demand that she tell me all she knows of Cerebus and his false miracles. And yet I am certain that it is only a series of coincidences which impede my progress, nothing more. Improbable events, but they are easily explainable within the confines of natural law. They are unexpected, true, but they are

My death, he described accurately. What seems a perverse coincidence, an inexplicable twist of fate; over-estimating the height of my horse; falling; being killed instantly, was, in actuality, the culmination of many moves on many chess-boards. An inevitability; a convergence of the necessary and the expedient in many realities and on many levels. I am convinced, in retrospect, that I would have succeeded. That the flourishing of freedom in the Feldwar Valley would have made the rapid corruption and disintegration of freedom in the Sepran Empire vivid by contrast, and would have brought about a popular uprising and a revival of our old ideals.

But such was not to be the case.

For it was my son, Alfred, who would lead the assault on Iest.

YOUR MOVE ...

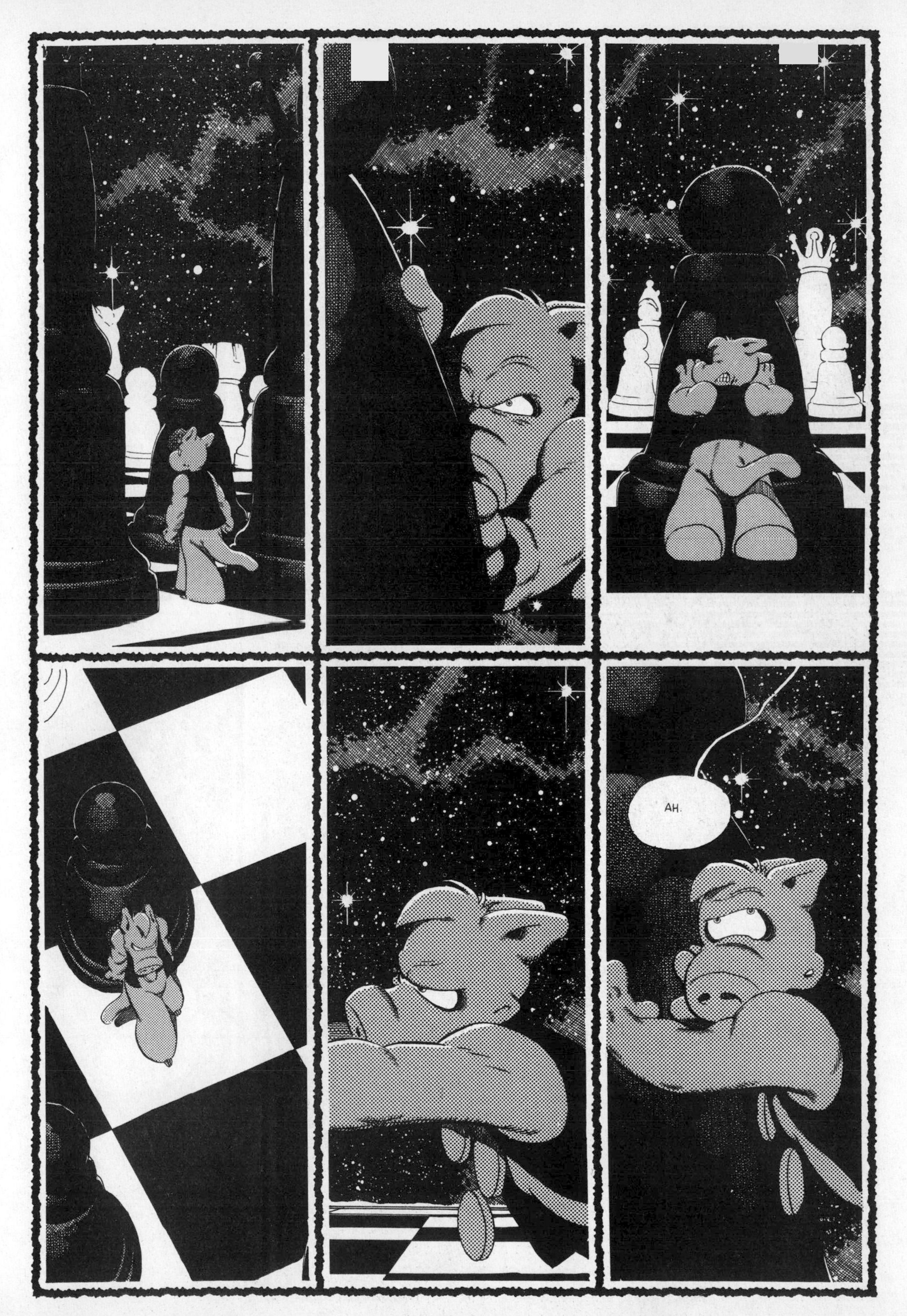
AH.

ANOTHER MYSTERIOUS APPEARANCE
THE FORMER PONTIFF OF THE EASTERN CHURCH -- BELIEVED LONG-DEAD...
A HIDDEN PRIEST CARD, THEN...

WELCOME.

WELCOME BACK TO IMESH.

WE SHOULD KILL ALL THE WOMEN...
NO NO
YES YES
WHAT IF THEY FORCE US TO HAVE ROACH SEX?
THEY WON'T
THEY WILL!
WE'LL WEAKEN AND MAKE THEM ROACH QUEEN ...
NO WE WON'T
YES WE WILL!
THEN WE'LL JUST KILL THE GOOD-LOOKING ONES
NO NO
YES YES
WE'LL JUST HAVE ROACH SEX WITH THE UGLY ONES ...
NO NO
YES YES
WHO WILL?
YOU WILL
WE'LL HAVE ROACH SEX AND THEN KILL THEM
FORGIVE ME, TARIM FOR I HAVE SINNED --I HAVE HAD THOUGHTS OF SLIDING MY SWEATY PALMS ALONG THE INNERTHIGHSOFA POUTINGADOLESCENT BLONDENYMPHET STROKINGANDPRODDING ANDBITINGHERFIRM UPTHRUSTHUMUNGOUS
SHUT UP SHUT UP
THEY'RE OUT THERE NOW ... MOIST AND
SHUT UP SHUT UP
WILLING AND
SHUT UP SHUT UP
AND LAS-VICIOUS
SHUT UP SHUT UP
BESIDES, ISN'T IT "LASCIVIOUS" ?

THE WAY PEOPLE HAVE BEEN APPEARING AND DISAPPEARING WE UNDOUBTEDLY HAVE VERY LITTLE TIME...
SO! I SHALL ASK MY QUESTIONS DIRECTLY...
WHEN THE BLACK TOWER DISAPPEARED ...
WHERE DID YOU GO?
THE MOON.

OF COURSE! OF COURSE! THE MOON!
CELESTIAL ASPECT OF THE GODDESS HERSELF
YOU ARE UNQUESTIONABLY A FAVOURITE OF HERS...
AS AM I...
TELL ME QUICKLY, THEN...
WHAT IS LIFE LIKE ON THE MOON?

LIKE IMESH.
eh?

EMPTY...
GRAY...
EXCEPT FOR ONE CRAZY OLD COOT CALLED THE JUDGE...
WHO TALKS ABOUT NOTHING BUT THE GODDESS ALL THE TIME...

HAHAHAHA

YES. OF COURSE OF COURSE ...
ALL OF US; HER HUMBLE AND LOYAL SERVANTS; LIVING IN SPLENDID ...

LOOK!

SHE MANIFESTS HERSELF!!

IT'S THE LIVING GODDESS ...
DO YOU SEE?
SHE
SHE LOOKS LIKE MY BELOVED SEDRA

HAHA
DO YOU SEE?

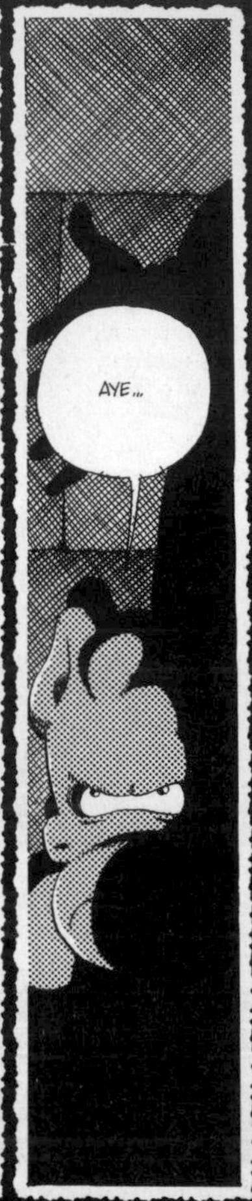
AYE...

OH GODDESS, GIVER OF ALL LIFE; K'COR YOUR HUMBLE SERVANT BIDS YOU GRANT MY FONDEST WISH-- I... I...

NO
PLEASE ...

PLEASE DON'T LEAVE ME ...

PLEASE...

Sedra left him; departing in the night with one of his slaves. He had entrusted her with the formula for his narcotic foodstuff, never once considering that the knowledge made her independent of the absolute control he wielded over Imesh's populace. She left a short note that was distant and impersonal. His heart and his spirit broken, he drove the entire population of the city out through a series of underground passage-ways which he then sealed with structural collapses. Now totally and completely alone, he dismantled the scaffolding around his half-completed monument with his bare hands over the course of several weeks. The monument was of great and vital significance on many of the inter-connecting chessboards; alignments of power and influence ebbed and flowed in its proximity. Its completion would have wrought profound and lasting change. Now it is a mere trinket; a curiosity for future generations to puzzle over and ultimately dismiss. Once a key figure and active piece, K'cor has sunk into dementia of the commonest sort; holding conversations with a Goddess for whom he is less than a joke; he will end his days broken and without significance. It is perhaps the oldest and most fundamental of the story that is Chess; the mighty King, reduced to a mere Pawn of his Queen. When he pointed out his manifestation to you, whom did you see?

Astoria.

You would do well to muse on that and keep it always close to your thoughts. Queen to King's Bishop Three.

My son Alfred was not an ignorant man, nor an evil man; but his motivation in taking both my place and my name were of a basic and mundane sort. Conquest was, to him, as it is to most Great Conquerors; a balm for his vanity, a jewel in his imagined crown. He thought himself clever that he chose to ignore the corrupt and insular matriarchy which ruled the Upper City. As I had timed and planned conquest for the right and perfect moment, it took place with clockwork efficiency and a minimal loss of life. Alfred sank happily into the opulent and coddled existence of the feted and legendary Military Hero. He grew fat on exotic foods and wines, his banquets were legendary and his appetites virtually without limit. Commerce flourished for a time — a short time, of course, as empires go, but a decade or so; which is always sufficient to convince the vainglorious and narrowly focussed that a Golden Age of their own devising has been wrought. The problem, of course, was that he had no thought or plan for any future; either his own or Iest's. The natural schism between matriarchy and patriarchy asserted itself in myriad ways in the day-to-day-life of the city state. Those subtle manifestations which interwove themselves through his government, clinging matriarchal vines, escaped Alfred's notice altogether. The more Noticeable and Assertive he blindly attempted to crush and curtail with Edicts, and when his Proclamations failed, with the brute force of his soldiers. Not surprisingly, the masses who had hailed his triumph a handful of years before, now turned on him. The Noticeable and Assertive rebellions vanished, but the subtle and insinuated tendrils multiplied exponentially; strangling the body politic and effectively neutralizing its now impotent leader.

Alfred retreated into Mysticism; surrounding himself with both Masters of Profound Wisdom and pitiful charlatans. As a latecomer to the hidden realms, now a corrupt and self-indulgent wastrel, he sought to purchase his knowledge and mystical power with mere gold. Naturally enough, his ignorance brought him power sufficient only to consume what was left of his failing health.

Those few friends remaining to him sickened mysteriously and died. Catastrophes swept the remnants of his empire.

Driven nearly mad by his precipitous fall from grace, realization came late to Alfred that the Great Forces which he had believed himself to Contain, he had merely Held. Further, that what he had Held was now slipping, like sand, between his fingers. He replaced his governors and his Statesmen with Conjurors and Mages. He declared that all who would follow him were to be called Illusionists. All Life was an Illusion, he said, which explained (to his satisfaction, at least), the mire of failure in which he found himself. He conferred his name on the circle of devotees which surrounded him, and empowered them to do the same.

By the time of his death at the age of forty-one, fully one third of the population of the Lower City was named Suenteus Po and believed themselves to function within a single, divine consciousness. Since that day, Illusionism has alternately flourished and declined across the length and breadth of Estarcion; rising and then falling, gaining influence and prominence one day, and being subjected to persecution and purge the next.

In my subsequent incarnation, I was born to a gold miner and his wife in Rivershire Province, twenty years after Alfred's death.

YOU INFORMED HER THAT I WISH TO SPEAK WITH HER.
YES, GREAT CIRIN...
AND?

SHE OUTLINED TWO...
...CONDITIONS
CONDITIONS!

YES, GREAT CIRIN--SHE WANTS TO MEET ON NEUTRAL GROUND.
THE FELDWAR SUITE AT THE REGENCY HOTEL ...

I SEE
AND WHAT ELSE?

SHE WANTS A NEW...
A NEW OUTFIT...
FOR THE OCCASION.

A NEW ...

GO BACK AND INFORM MISS HIGH AND MIGHTY THAT SHE IS MY PRISONER
NOT I HERS

I SAID AS MUCH, GREAT CIRIN-- PRACTICALLY THOSE VERY WORDS ...

AND WHAT DID SHE SAY?

uh
SHE--

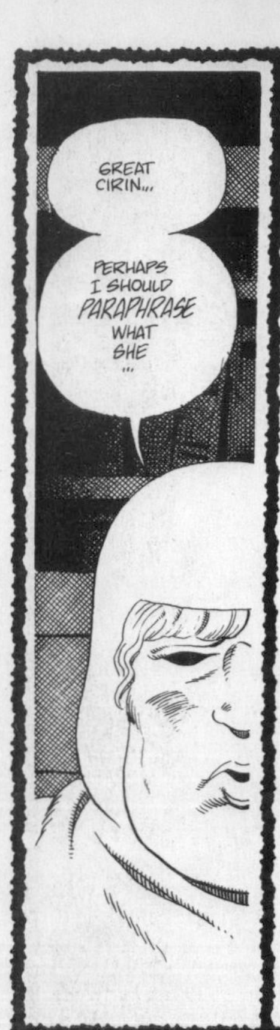
GREAT CIRIN...
PERHAPS I SHOULD PARAPHRASE WHAT SHE ...

HER EXACT WORDS...

SHE SAID
"TELL THE OLD"
"BATTLE"
"AXE"
" THAT IF SHE'S COMING " uh
"COMING CRAWLING"
"TO ME"

" THAT SHE MUST uh MUST HAVE"
"BOTH"
uh
"TITS"
uh
"CAUGHT"
uh
"CAUGHT"
" IN THE"
uh
" WRINGER

SMASH
SMASH
SMASH

TAKE HER FROM HER CELL TOMORROW MORNING
HAVE HER CLEANED UP...
THEN TAKE HER...

SHOPPING...

BUY HER WHATEVER SHE *WANTS* ...

AND HAVE HER AT THE *REGENCY* NO LATER THAN *NOON*...

YES, GREAT CIRIN...

YOUR MOVE
MOST HOLY!

WITHIN THE HOUR: NOW FULLY-ARMED, THE PICTS STREAM FROM HUNDREDS OF HIDDEN ENTRANCES TO THEIR UNDERGROUND LAND -- THEIR TORCHES BLAZE TO LIFE IN THE CHILL AIR OF EARLY WINTER IN THE RED MARCHES... A LIGHT, DAMP BREEZE AT THEIR BACKS, THE GRIM FIGURES MOVE EASILY AND SURE-FOOTEDLY, SLOWLY ARRANGING THEMSELVES INTO ORDERED UNITS; TWELVE MEN DEEP AND TWELVE MEN ABREAST... AS THE LAST FEW MEN TAKE THEIR PLACES....

A CRY RISES UP FROM THE THROATS OF EACH MEMBER -- AND THE FORMATION LAUNCHES ITSELF FORWARD INTO THE NIGHT, TORCHES THRUST FORWARD AND UP IN THE DIRECTION OF THE PICTS ANCIENT AND ANCESTRAL HOME; IEST!

OH TARIM I CAN SEE THE PERFECTION OF YOUR MULTIVERSE FROM THE SMALLEST INSECT TO THE MIGHTIEST AND MOST PERFECT OF YOUR PROPHETS (ME)
BUT I AM SORELY ROACH-VEXED GREAT TARIM
I KNOW NOT WHAT I DO...
I-- I--
I'M JUST A BALL OF CONFUSION--THAT'S WHAT THE ROACH IS TODAY...
HEY HEY
I WANT TO REMAIN PURE IN ROACH-BODY AND ROACH-MIND BUT THE BOTTOM PART OF MY ROACH-SKULL CHEST EMBLEM™ IS...
FEEL LIKE I'M A HUNKAHUNKAHUNKA BURNING UP...
IS PINK AND WET AND SLIMY AND..
AND..
SHUT UP
SHUT UP
SHUT UP
SHUT UP
SHUT UP
SHUT UP
SHUT UP
SHUT UP
SHUT UP
MUST... THINK... OF SOMETHING... ELSE.
BUT WHAT?
WHAT? WHAT?
WHAT?
WHAT?
MUST RETELL MY ORIGIN! YES!!
THE DEADLY GAME OF CHESS CONTINUES! SUENTEUS PO'S NEXT MOVE IS KING'S BISHOP TO QUEEN'S BISHOP FOUR...
DISASTER LOOMS!
HOW VERY LIKE DISASTER ...
NONO! YESYES
NONO...
NONO!
ONCE A MILD-MANNERED MERCHANT-ROACH OF NOT INSUBSTANTIAL MEANS, IN A FREAK LABORATORY ACCIDENT, I WAS BITTEN BY A RADIOACTIVE NYMPHOMANIAC...
HARHAR
SNORTSNORT
NONO!
YESYES

A CRISIS OF FAITH! UNCERTAINTY AT EVERY LEVEL OF THE MULTIVERSE -- EVEN THOSE PARTS WHICH HAVE BEEN REZONED FOR COMMERCIAL USE
ONLY ONE THING CAN SAVE US NOW
A SIGN OF THE DIVINE WILL OF THE LIVING TARIM -- PLEASE GREAT TARIM A SIGN --
JUST...
ONE
TEENSY
WEENSY
SIGN
A SIGN OF YOUR DIVINE WILL...
RENDER UNTO ME, YOUR HUMBLE SERVANT
I DOUBT YOU NOT BUT HEAR ME, OH TARIM -- JUST FOR THE SAKE OF ARGUMENT ...
OH GREAT AND MERCIFUL TARIM -- I, THE MOST PERFECT OF YOUR PROPHETS, AM VERILY COMING APART AT THE ROACH SEAMS ...
NO NO
YES YES YES YES
NO NO NO NO
THAT'S NOT IT...
SOB
WHAT?
WHAT IS IT THEN?
MY... ORIGIN...
MUSTN'T... FORGET... MY ORIGIN...
ONCE A MILD-MANNERED MERCHANT OF LABORATORY SUPPLIES I ACCIDENTALLY LICKED A FREAK NYMPHOMANIAC...

AH HAVE ... I SAY
AH HAVE ARRIVED IN THE PROMISED ROACH LAND!
"JAW" "REE" "VEY" FOR THE BILINGUALLY INCLINED
LOWER FELDAN, THAT IS ...
IT'S THE DAWNING OF THE AGE OF THE BLATTIDAE AND THE PHYLLODROMIDAE
BRING FORTH YO' TIRED -- YO' POOR, YO' HUDDLED HARLOTS YEARNIN' FO' IMPREGNATION AND A ---
!
WELL SHUCK MAH CARAPACE AN' STIR-FRY ME IN THE SUNSHINE!!
PUNISHEROACH!! WHOSE MUST-HAVE DOUBLE-BAG FIRST APPEARANCE AND COMING--
(IF YOU'LL PARDON THE EXPRESSION)
-- WAS FORETOLD TO ME..
OH GREAT AND MERCIFUL TARIM...
PLEASE...
LET ME REPHRASE THAT....

One of the gold coins suddenly blazes with life again, pulsing with an inner radiance and hurling from it a shower of golden particles which appear to swim upward through the murky waters in random patterns; weaving, co-mingling. Then vanishing as quickly as they had appeared.

YO... YOU'VE COME BACK!
LOOK!
IT'S H-HIM!
OS-OSCAR O-OSCAR L-LOOK!
DO YOU SEE?
DO YOU SEE?!
POSEY! NO! NO!

POSEY--
QUIET
HAVE YOU LOST YOUR --
WE'RE SUH-SAVED OSCAR...
WE'RE S-SUH-SAVED ...
HALT

PRISONER!
YOU HAVE R-RETURNED TO F-FUH-FULFILL THE P-P-PUH-PROPHECIES ...

YOU HAVE COME TO SUH-STRIKE THE SH-SHACKLES FUH-FROM OUR L-L-LIMBS
TO S-SMITE OUR OPUH-OPPRESSORS

UNH!

I N-N-NUH NEVER ...

STOPPED...
B-BUH-BELIEVING IN...

WHIPP

Y-YOU

WHIPP

MOVE ALONG.

The hardest lesson for the Reformer to learn is that he chooses not whom he inspires, nor does he choose the form and the substance of that inspiration. To you, Archbishop Posey was a mere functionary of the Eastern Church and since your first meeting at the time of your investiture as Pontiff, you have scarcely given his existence a second thought. But it was he who was the most profoundly affected by you. You came to him in a dream some weeks ago and beckoned to him. A life-long coward, his belief in you as an incarnation of the Living Tarim overcame his fear. He fled the Sequester of the Church in the Upper City to search for you, praying fervently to Tarim to guide him in his quest, to be by your side, whether in victory or in martyrdom. He was arrested half a block from Dino's Cafe and sentenced on the spot to two years of hard labour. Against all reasonable expectations, he had survived these weeks of incarceration, his body bruised and weighted by exhaustion, but his spirit unbroken; unshakable in his conviction that you were the Redeemer and that you would return and fulfill the prophecies.

He died happily and at peace.

Because of you.

At an early age, I evinced interest in the primitive gold-smithing tools in my father's small workshop. Ours was a small and impoverished village and there was little gold to practice on except the gold coins which served, as they do to this day, as the foundation of each family income in Iest; whether entrusted to a patriarch in the Lower City or to a matriarch in the Upper City. One side of the coin is struck with the emblem of the local governor at the time it is issued. The other side is blank and is traditionally carved with a symbol of the family who possesses it. I grew adept at carving these symbols, each more elaborate than the last and soon families were coming from neighbouring villages to have my design added to their coins. I learned to scrape traces of gold from the coin to melt and use as inlay. I invented new tools for engraving finer lines and patterns. A devout Tarimite, I carved His Name in ancient Pigtish rune letters on each coin brought to me. Healers and apothecaries began using the coins in their treatment of family ailments. Miracles were spoken of in hushed whispers and soon more people came from the larger villages and then from the City itself; nobles, lawyers and merchants. As whispers became words and words became legend, the coins seemed imbued with the belief of the people made manifest. Priests of the Eastern Church grew jealous of this faith; felt their influence and control of the population waning with each passing day in Rivershire Province and elsewhere. They took their case to Suenteus Po III, informing him that I was building a kingdom within Iest and that I had declared my family carvings as more worthy of loyalty than the governors' emblems on the obverse.

When they came to arrest me, I knew this had all transpired before many times; as if I was an actor in a play. I remembered my life as Suenteus Po I. As they read the indictment I had the curious sensation that I was imprisoning myself.

As they led me away, there was a flurry of resistance and some blood was spilt. I told my defenders to stand back; not to make it any worse than it already was. They complied and I could see from their expressions that they, too, now felt like actors in a play.

Many of them wept openly.

I was brought before Suenteus Po III, a bloated caricature of his father; my son, Alfred. He was amused by my threadbare appearance, my regional dialect. The mages and charlatans who held posts in his Illusionist court were amused as well; they felt that the priests of Tarim had finally lost their minds, seeing a threat in this misshapen peasant. Their questions had a comic turn to them. I answered each question as simply and as honestly as I could. I felt I was part of the joke and that soon they would tire of me as an amusement and I would be set free; though what my fate would have been, then, I have no idea.

But, Po III was seized with the thought of enlargening the jest. He asked if I carved only coins that were minted in the Lower City. No, I said, a number of them had been brought to me by representatives of the Great Ladies of the Upper City. He asked if I felt no qualms as a good Tarimite in crafting coins with the Goddess on them. I admitted that it had troubled me, but I felt that the Word of Tarim should be given freely everywhere and paid witness to. He turned then to his senior advisors and asked what they thought Great Andrena, leader of the Council of the Goddess in the Upper City might think of my heresy. There was much levity at the very prospect.

The play had resumed its course.

He asked if I had any further words for the illustrious body before me. I said nothing. He informed me that arrangements would be made to transfer me to the custody of the Guardians of the Upper City. With a theatrical gesture, he drew forth a small bowl of scented water and dipped his fingers lightly into it.

I wash my hands of you, he said, and I was led away.

That night I was led before Great Andrena. There is no need for me to relate our conversation; the course of the Trial. You experienced it yourself when you tried Astoria as Pope.

You didn't sentence her to death. Events intervened and you ascended instead.

This gives us small cause for hope, does it not?

I was taken to a courtyard in a small prison attached to the Council Building. The charges against me were read again and then the fagots were lighted and I was burned as a heretic.

My experience taught me that there is no benefit and little wisdom in attempting to influence the minds and the wills of the mass of people. In both my lives I have described to you, I sought that kind of influence and effect; I was a Reformer. In my succeeding lives, I have seen the long-range effects that profound change always brings about. Each Great Movement is sown with the seeds of its own destruction; its corruption and decay as inevitable as Death itself. In each succeeding life I've led, after leaving my parents' house, I have sought a simple and uneventful existence. My quarters are always mean and rudimentary; a bed where I might sleep, a table where I might eat and a chair for sitting on. At this moment, I live in a one room apartment in Iest's Lower City. I have no friends and no contact with any of my relatives. My one luxury is a crude chessboard, made from a discarded packing crate; the pieces carved by hand from scraps of firewood.

Queen's bishop to King's Bishop Four.

Its flourish of energy
spent, it becomes dull
and lustreless once more.
One side of its companion
coin flares with radiance,
issuing forth nine perfect
spheres of white light; one
tiny, three slightly larger,
one immense and heavy,
one compressed and nearly
transparent, two smaller
than these and of compar-
able size and then another
so small as to be nearly
invisible. The golden light
vanishes and all is as it
had been.

IT LOOKS LIKE OLD ELROD GOT HYAR NOT A MOMENT TOO SOON!
YOU LOOK ah say-- YOU LOOK LIKE A POSTER BOY FO' TH' ULCER OF TH' MONTH CLUB!
NOT LISTENING
NOT NO
'STRESS ROACH' THAT IS...
WHAT Y'ALL NEED IS A LITTLE 'ARE' AN' 'ARE'
NO LISTENING
NO
'RUTTIN' AN' RELAXATION'
NO NOT NO NO NOT
YESSIR! NOTHIN' SAYS LOVIN' LIKE A SELF-LUBRICATIN' OVEN!
TUNA PIE, THAT IS...
SO WHUT SAY WE SCOOP OUR-SELVES A COUPLE O'HUNDRED POUNDS O'ROACHLAND'S GRADE DOUBLE 'A' PRIME BIMBO FLESH--
NO
NOT LISTENING NOT
AN' HAVE US A HORIZONTAL RODEO!
LISTENING
LISTENING...
SPIT IT OUT-- SON-- WHUT HAVE Y'GOT T'SAY F' Y'SELF?
NOT NOT NOT!
LISTENING!
NOT NOT NOT
SPEAK ah say SPEAK UP, SON!
EVERYTHIN'S TWITCHIN' BUT NOTHIN'S MAKIN' IT PAST Y' LARYNX!

I-- CAN'T

OH HO!

NOT ENOUGH ah say-- NOT ENOUGH STARCH IN YO' LINGUINI, HUH?

WELL, Y'ALL JEST LEAVE IT TO DOCTOR ELROD...
AH'VE GOT JEST TH' THANG ...
RAGHT HYAR

FILTHY-- ah say FILTHY LOWER FELDAN POST CARDS!

OH NO
HM.
"WHAT THE MAID SAW"
"NAUGHTY DEBUTANTE"
AH KNOW IT'S HYAR SOMEPLACE ...

AHHA!
CUCUMBER QUEEN!
SPLURT

SAY! WHO ARE YOU?
HOW DID YOU GET IN HERE
AND WHY AREN'T YOU BETTER-LOOKING?
NO ONE CAN STOP THE EYE IN THE PYRAMID LORD JULIUS ...
WE ARE AS SMOKE
WE ARE YOUR SHADOW
WE ARE EVERYWHERE

HMM YOU'RE LIKE SMOKE -- YOU'RE MY SHADOW AND YOU'RE EVERYWHERE

ARE YOU BIGGER THAN A BREAD BOX?

WE ARE THE HAMMER OF VENGEANCE
WE ARE THE INSTRUMENT OF THE GODDESS' DIVINE RETRIBUTION
WE WILL CRUSH YOUR BLOATED AND LEECHING CITY-STATE LIKE AN INSECT....

OH! SO YOU ARE BIGGER THAN A BREAD BOX...

YOUR END IS NEAR, LORD JULIUS ...

REALLY?
I WAS JUST THINKING IT'S QUITE A BIT CLOSER TO THE FLOOR THAN IT USED TO BE...

YOU DISAPPOINT ME.

WELL, I'M NOT VERY HAPPY ABOUT IT, EITHER, BUT THAT'S GRAVITY FOR YOU

I EXPECTED MORE FIGHT, MORE SPIRIT FROM THE FABLED GRANDLORD OF PALNU

YOU DID?

SAY! MAYBE THAT PUBLICIST WASN'T SUCH A WASTE OF PIN MONEY AFTER ALL

DEFEND YOURSELF!

WELL, I'M NOT A HALF-BAD DANCER ... AND IN THE RIGHT EVENING-WEAR I'VE BEEN KNOWN TO LOOK POSITIVELY ...

SLASH

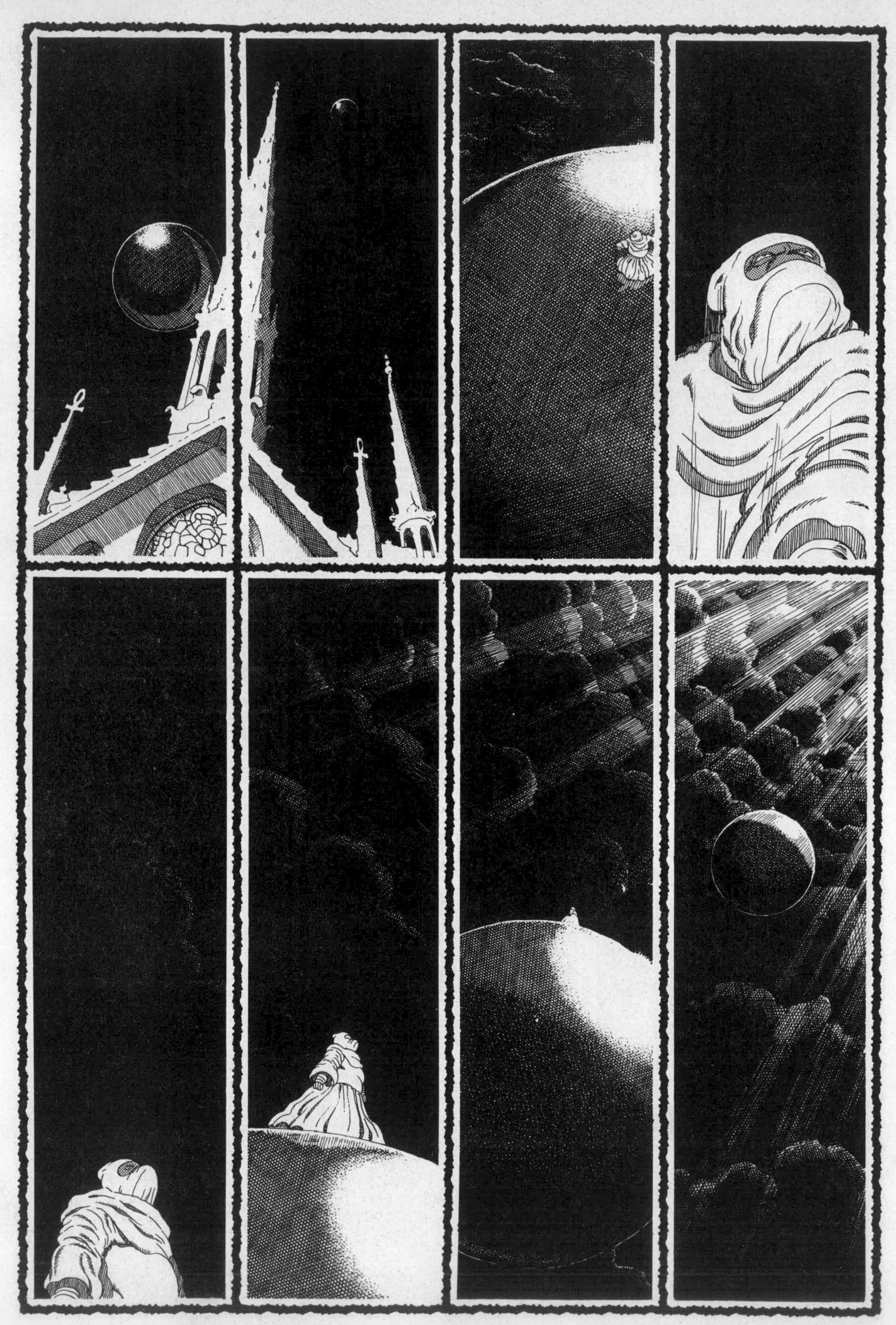

OH GREAT TERIM
GODDESS OF ALL--

AH HATE THIS CITY ...
AH HATE W'ARIN' THIS SILLY-ASS BOOWEY TIE ...
AN' AH HATE FEELIN' BREEZES ON TH' BACK O' MUH NECK ...
AH HATE SHAVIN' EVER' DAMN DAY...
YEW KNOW WHUD ELTS AH HATE ?
YEW HATE HAVIN' YER GOD-GIVEN RIGHT T'CALL A NO-ACCOUNT FATASS SOW-BELLY BITCH (WHO'S GITTIN' IN YER FACE) A NO-ACCOUNT FATASS SOW-BELLY BITCH, TAKEN AWAY FRUM YOU BY SUM NO-ACCOUNT, FATASS SOW-BELLY BITCH...

Y'FERGOT TH' "MUFF-DIVIN'" PART, AGIN.
HOW MINNY DAYS'VE AH SARVED O'MY PROBAYSHUNEERY PARIOD?
TWO.
AN' HOW MINNY DAYS'VE AH GOT LEFT AFORE MUH GOOD BUH-HAYVYER HEARIN'?
EIGHDY-SEVEN.
AH HATE THIS CITY ...
LAST CALL!

YOUR MOVE

The first coin decomposes
into a blackened, tarry mass.
The other becomes corrupt
and dark along its one edge.

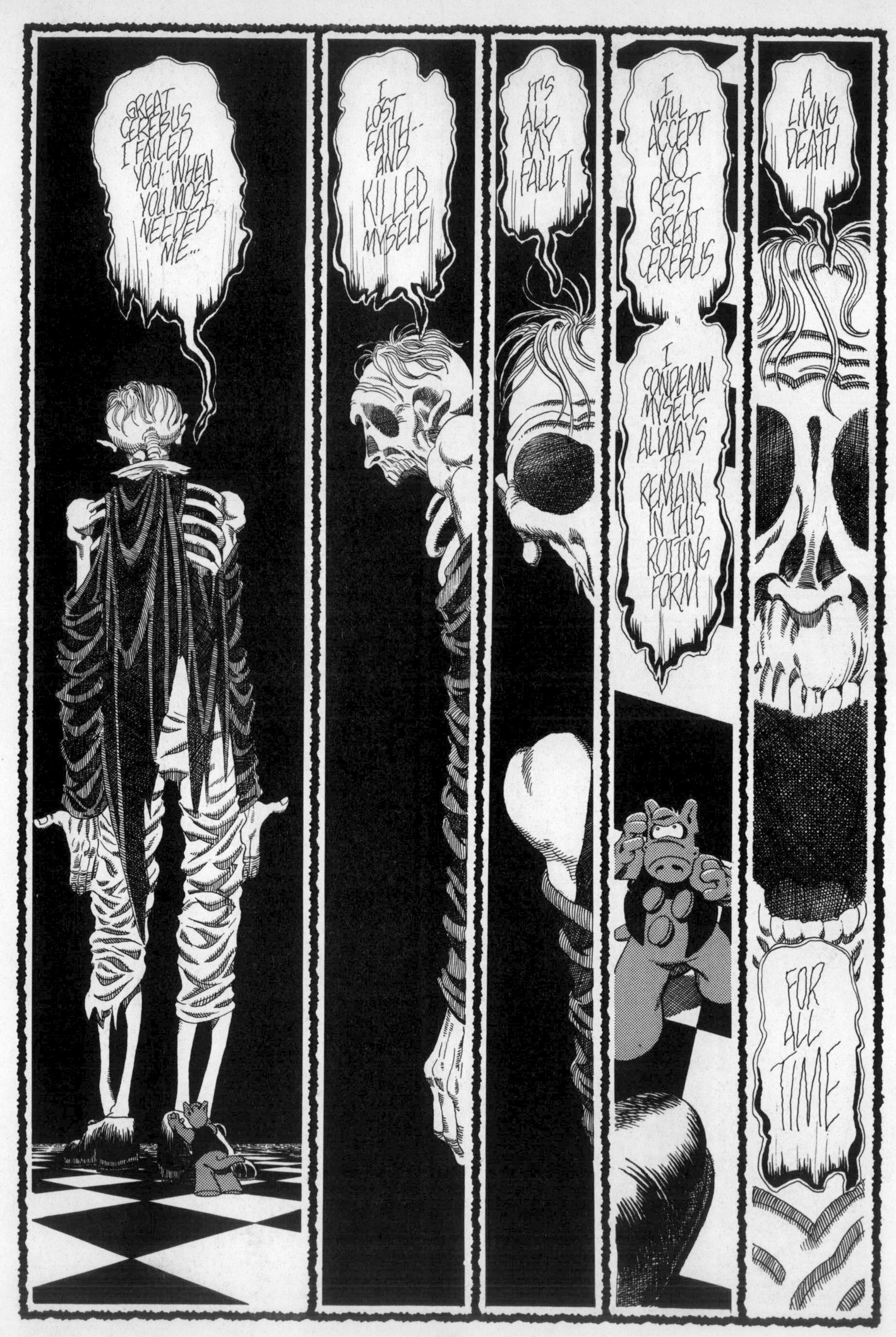
GREAT CEREBUS I FAILED YOU--WHEN YOU MOST NEEDED ME...
I LOST FAITH-- AND KILLED MYSELF
IT'S ALL MY FAULT
I WILL ACCEPT NO REST GREAT CEREBUS
I CONDEMN MYSELF ALWAYS TO REMAIN IN THIS ROTTING FORM
A LIVING DEATH
FOR ALL TIME

Our match draws to a close, which is for the best, as I have violated my self-imposed guide-line that I would discuss with you only my own lives and have succumbed, instead, to the temptation to pass several subjective judgements.

We begin by criticizing ourselves and end by criticizing others.

Your welcoming of an alliance with Bran Mak Mufin is what sealed your fate as Eastern Pontiff. Throughout his myriad previous lives, his pattern has remained, unfortunately, consistent; marked by a rise to prominence, influence and leadership in the environment in which he finds himself; followed by the consolidation of that authority into a reign which engenders nearly absolute fealty. His ensuing conversion to another leader's cause and standard brings about the consequent fracturing of his own constituency whose loyalty, freely given, is to him alone and not some manner of spiritual currency, as is his misunderstanding, to be spent as he deems fit. The resultant isolation from his followers is punctuated by a swift decline into sycophancy, idolatry and rationalisation. In the end, there comes an Event of Crisis which precipitates disillusionment, despair and ritualistic suicide. In its turn, this brings about the Fall of the Object of his new and abiding faith. At each conclusion, as a kind of penance, peculiar to himself, he resolves that his spirit will remain forever earth-bound in its corpus of the moment. The Grand Folly in this, of course, is that the decay and decomposition of his physical form is as inevitable as the onset of the seasons.

And so, his spirit migrates endlessly in place, resisting any opportunity to rise; to advance. He is a tadpole, whose fear of becoming a frog is as grandly ridiculous as it is contrary to natural law; as all-encompassing as it is willfully destructive.

His last conscious awareness, just now, of himself and his spirit's long history, was a mental image of disconnected vertebrae, stretching out into Infinity; each fractured and fragmented; a series of failures of will and impediments to Progress consisting in equal measure, of Fear and Doubt and Betrayal.

This Awareness is new to him and it is not altogether unlikely that a permanent madness will be the result; and that, in his lives yet-to-be, the destructive pattern which has asserted itself, might establish itself more firmly, or that his destructive nature might be amplified in some unforseen manner.

Madness, like Progress is only one of the many choices available to the human spirit in its migration through many centuries and many lives.

But now, he is truly the Dark Bishop of your choosing.

But with New Awareness, all things become possible.

So this, too, gives us small cause for hope, does it not?

MR. HAMMOND THIS IS MOST INAPPROPRIATE. IT IS THE MIDDLE OF THE NIGHT...
I UNDERSTAND -- BUT I MUST SEE MRS. THATCHER
IMMEDIATELY...
MRS. THATCHER IS ASLEEP. IN THE MIDDLE OF THE NIGHT, SLEEP IS THE ONLY APPROPRIATE BEHAVIOUR...
GO HOME, MR. HAMMOND. TOMORROW YOU CAN...

IT CONCERNS THE ASCENSION

I'VE JUST COME FROM THE CONSTRUCTION SITE...
SOMETHING'S
SOMETHING'S NOT RIGHT.

COME IN, MR. HAMMOND
YES...
YES. THANK YOU.

SUDDENLY, GALE-FORCE WINDS OUT OF THE EAST LASH AT THE PIGT WAR PARTY... BALLS OF LIGHTNING ROLL WITH UNNATURAL SLOWNESS THROUGH THE GRAY MASS OF CLOUDS ABOVE, BILLOWING, THEN ODDLY FROZEN IN EERIE PATTERNS, THROWING OFF ROPES OF BLUE AND YELLOW ELECTRICAL FORCE WHICH ARC ACROSS THE LENGTH AND BREADTH OF THE SKY. AND THEN THE SKY FLARES BRIGHTLY, THE COLOUR OF BLEACHED BONES AND IS FILLED, INSTANTLY, WITH ENORMOUS INSECT CREATURES, WHOSE MANIACAL SHRIEKING DROWNS OUT EVEN THE HOWL OF THE RISING WINDS. THE DEMON-BEINGS
VANISH AS SWIFTLY AS THEY HAD APPEARED. THE PUNISHING WINDS SUBSIDE WITH THEIR PASSING AND THE SKY IS FILLED, ONCE MORE, WITH BLACK AND TERRIBLE CLOUD MASSES..... IT BEGINS TO RAIN; LARGE PELTING DROPS WHICH BECOME PELLETS OF HAIL. THE PELLETS, WITHIN MOMENTS, BECOME STONES, SOME THE SIZE OF A MAN'S FIST, LEAVING RED AND PURPLE WELTS ON EXPOSED FLESH...

CLARINDA...
WHAT IN
VERY SORRY TO DISTURB YOU, MRS. THATCHER ...
BUT MR. HAMMOND IS WAITING DOWNSTAIRS ...
HE MUST SPEAK WITH YOU RIGHT AWAY...
NONSENSE...
HAVE HIM COME BACK IN THE
SOMETHING'S GONE WRONG WITH THE ASCENSION.
I'LL BE DOWNSTAIRS AS SOON AS I'M DRESSED...

LISTEN.. CAN'T WE TALK THIS OVER?
MAYBE I COULD BUILD YOU GIRLS A NEW CLUBHOUSE OR SOMETHING...
THE TIME FOR TALK IS THROUGH!
REALLY? BY MY WATCH IT'S BARELY QUARTER PAST THE TIME FOR EMPTY THREATS ...

IT IS TIME FOR YOU TO DIE!
DIE?
WHAT ABOUT THE TIME FOR SNIVELLING, WHINING AND BEGGING FOR MERCY?
I KNEW I SHOULD'VE BOOKED AHEAD...

SLASH

I WILL MAKE YOUR DEMISE SLOW... LINGERING ...AND PAINFUL!

YOU'RE GOING TO MARRY ME?!
YOU FIEND!
SLASH

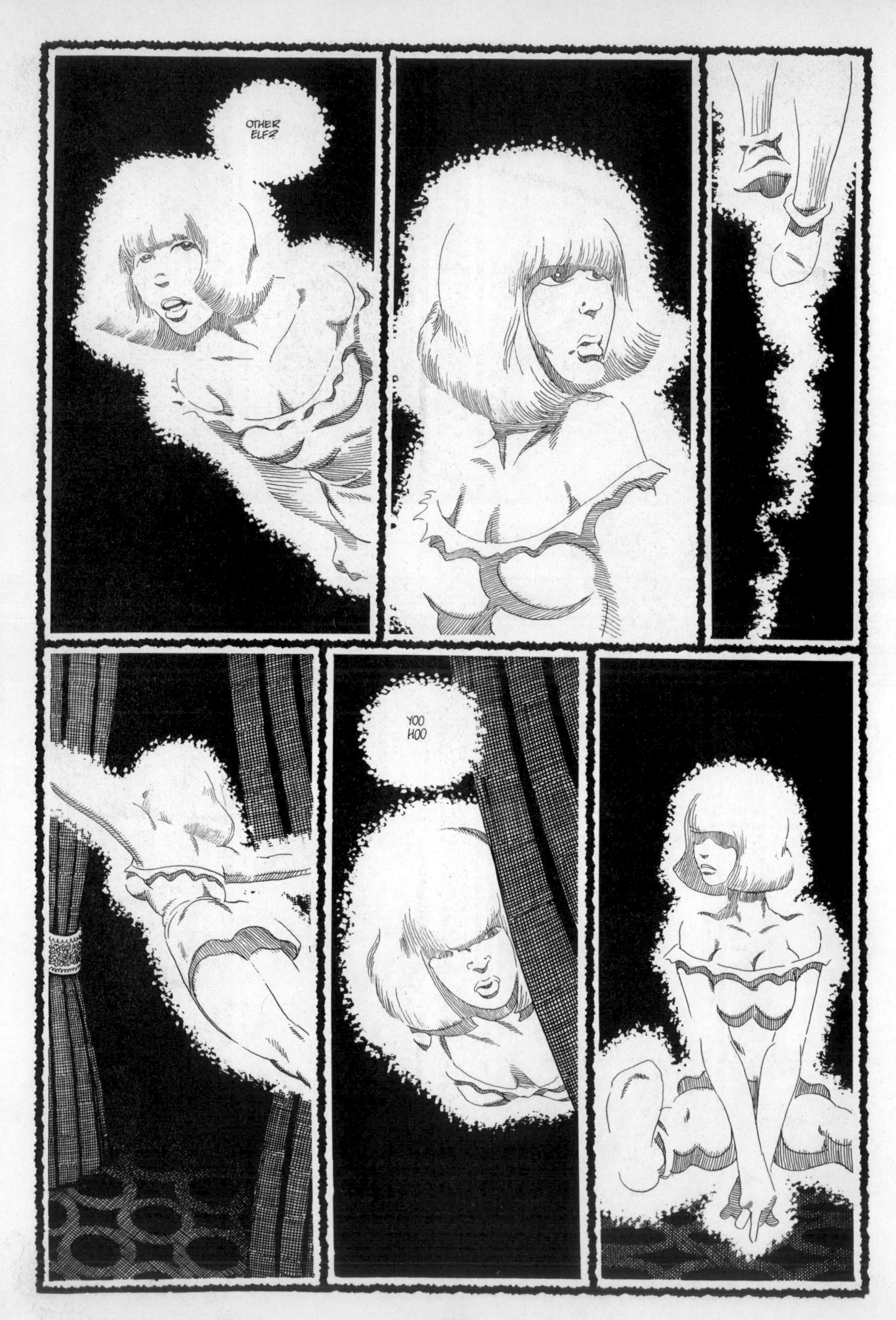
OTHER ELF?
YOO HOO

HSST.
ASTORIA
HSST!
ASTORIA!

ASTORIA.
THE WORD OF YOUR IMMINENT RELEASE IS BEING DISSEMINATED
BUT I'M ...
I'M AFRAID MANY OF THE FOLLOWERS ARE...
ARE ALREADY ...
OUT OF CONTROL
OF COURSE THEY ARE...
THAT'S THE WHOLE IDEA
LOOK--AH SAY LOOK AT IT THIS WAY, SON...
YOU KNOW THAT YOU'RE THE PUREST PROPHET OF TARIM'S WILL
AN AH KNOW YOU'RE THE PUREST PROPHET OF TARIM'S WILL
BUT IF--AH SAY--IF Y'ALL DON'T TEST IT, THEN HOW'S TARIM SUPPOSED TO KNOW?
NO
NO
NO
NO
YES
YES
YES
YES
YES

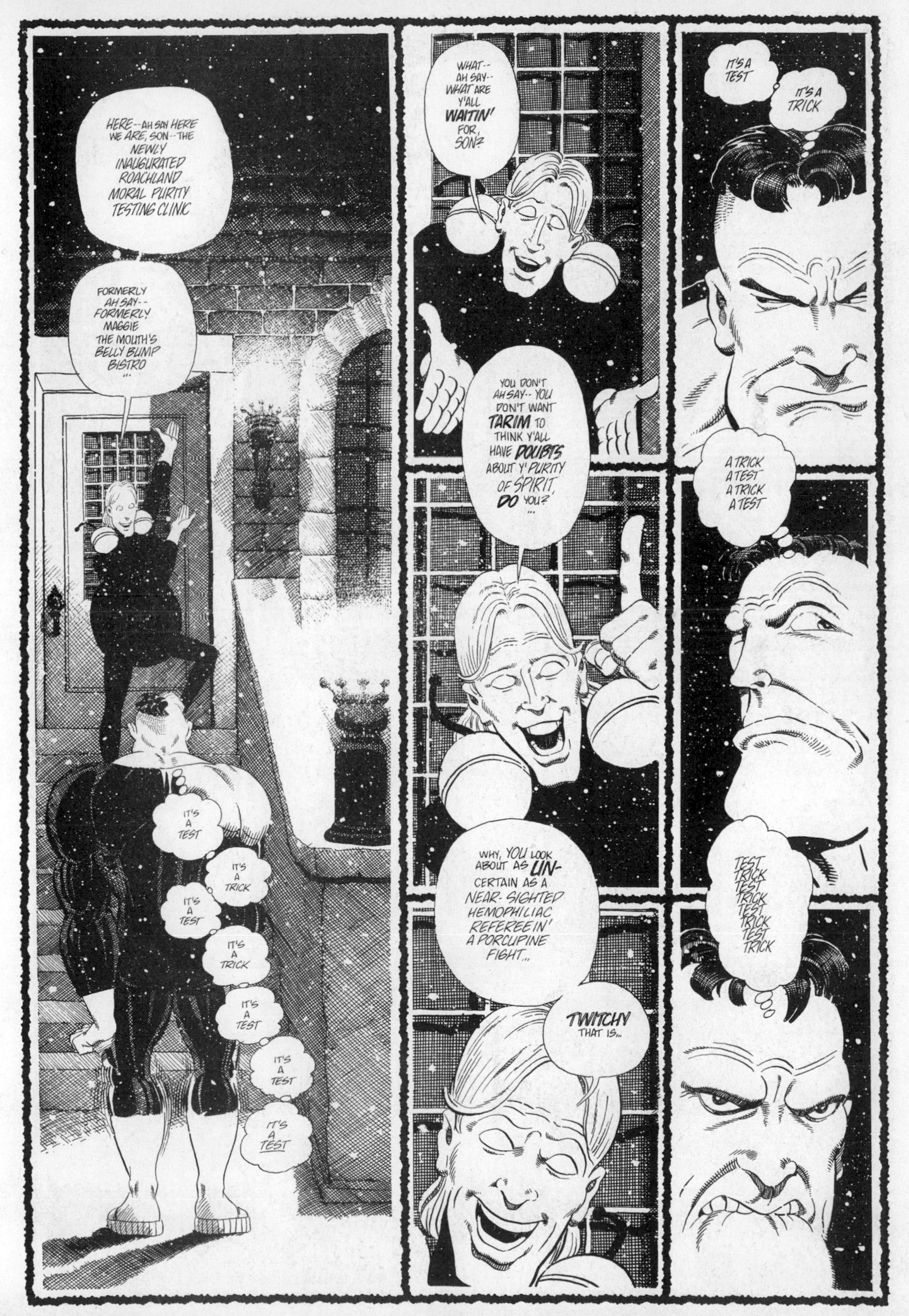
HERE--AH SAY HERE WE ARE, SON--THE NEWLY INAUGURATED ROACHLAND MORAL PURITY TESTING CLINIC
FORMERLY AH SAY--FORMERLY MAGGIE THE MOUTH'S BELLY BUMP BISTRO ...
IT'S A TEST
IT'S A TRICK
IT'S A TEST
IT'S A TRICK
IT'S A TEST
IT'S A TEST
IT'S A TEST
WHAT--AH SAY--WHAT ARE Y'ALL WAITIN' FOR, SON?
IT'S A TEST
IT'S A TRICK
YOU DON'T AH SAY--YOU DON'T WANT TARIM TO THINK Y'ALL HAVE DOUBTS ABOUT Y' PURITY OF SPIRIT, DO YOU? ...
A TRICK A TEST A TRICK A TEST
WHY, YOU LOOK ABOUT AS UN-CERTAIN AS A NEAR-SIGHTED HEMOPHILIAC REFEREE IN' A PORCUPINE FIGHT...
TEST TRICK TEST TRICK TEST TRICK TEST TRICK
TWITCHY THAT IS...

I TULD YEW SEM ASS I TULD EFRYWAN-- I SET--
LEST
COLE
WAL, AH DIDN'T HEAR YUH, VARMINT! AN' AH WANT ME ANOTHER DAMN BEER!!
I KENT DO EET...
LEST COLE EES
LEST
COLE
-- GIMME ANOTHER BEER AFORE AH TIE THIS HERE BAR STOOLIE AROUND YER SCRAWNY NECK
AH-AH-AH
REMEMBAIR YEWR PREWBATION PEERIODD ...
AH HATE THIS CITY....
Y' BETTER BRING A DAMP CLOTH UP WITH YUH--SUMBIDDY JEST PUKED ALL OVER YER ARMY COT--

FAR BE IT FROM ME TO KNOWINGLY CAST ASPERSIONS ON MY "KNOW-IT-ALL-WISE-GUY" EVIL TWIN IN ABSENTIA...
HOWEVER!
UNDER THE GENERALLY ACCEPTED TERMS AND PROVISIONS OF THE NEARLY UNIVERSAL LAWS WHICH GOVERN FORENSIC DEBATE...
I DON'T THINK IT'S TOO MUCH TO ASK THAT YOU REAPPEAR AT LEAST LONG ENOUGH FOR ONE, GOOD "I TOLD YOU SO" ...

SO, IN A WAY, MR. HAMMOND THIS IS EKTUALLY QUITE GOOD NEWS...
YES-- I
ISN'T IT.
YES-- YOU SEE, I THOUGHT IF...
OF COSS IT ALSO REPRESENTS A RRADICAL CHANGE--DOES IT NOT.
YES. WHEN I SAW THE
MMM?
YES. QUITE RADICAL.
YAAS.
AND GREAT CIRIN WAS MOST EMPHATIC THAT SHE BE KEPT ABREAST OF ENNY CHANGE ...
WELL, THAT WAS WHAT I THOUGHT WHEN I
ENNY CHANGE HOWEVER SLIGHT...
YES-- I...uh
YES.
CLARINDA?
YES, MRS. THATCHER
SIP
I'D LIKE A SMOLE GLOSS OF PORT, PLEASE.
YES, MRS. THATCHER
SIP
AND A CIGARETTE.
YES, MRS. THATCHER

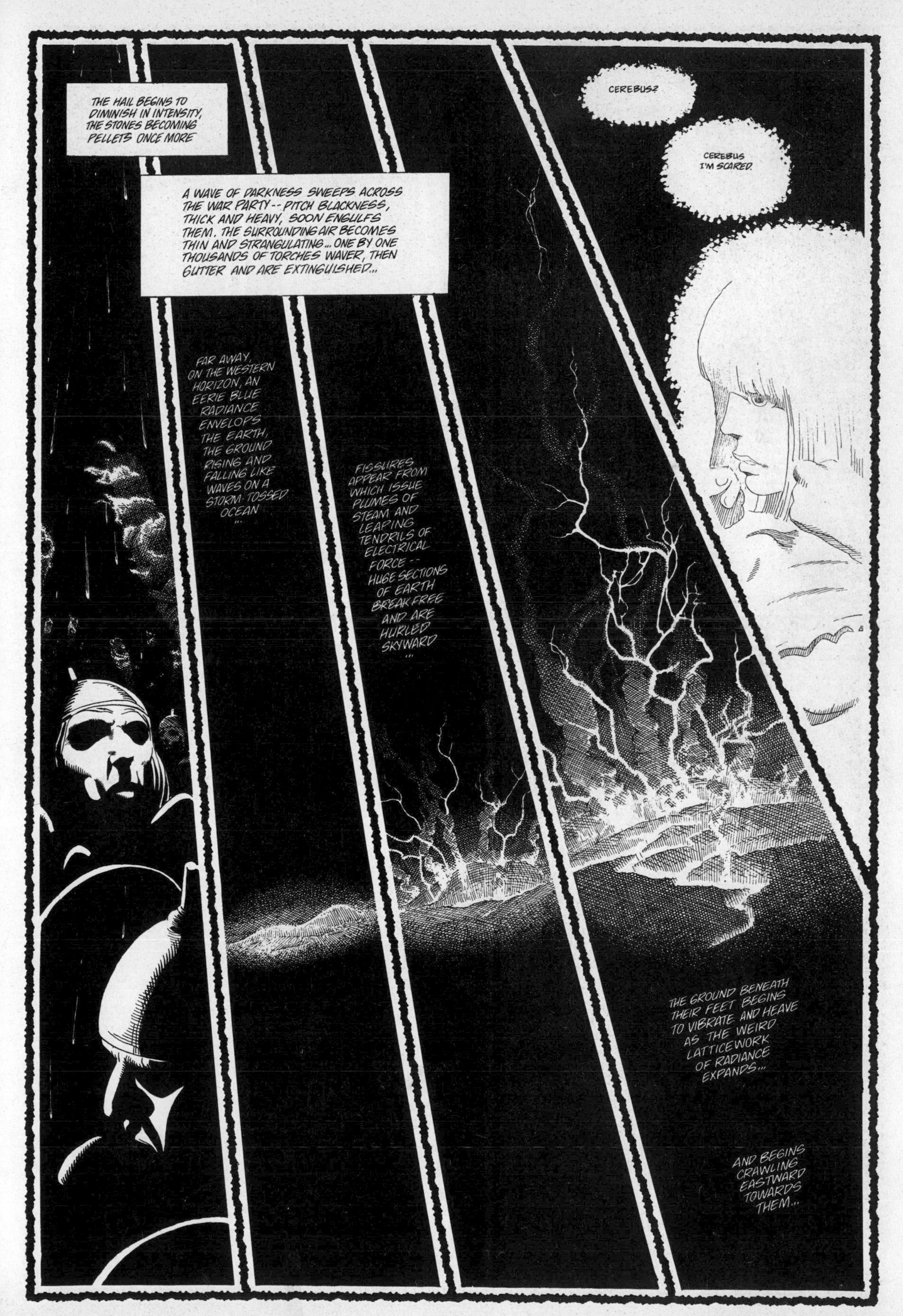
THE HAIL BEGINS TO DIMINISH IN INTENSITY, THE STONES BECOMING PELLETS ONCE MORE
A WAVE OF DARKNESS SWEEPS ACROSS THE WAR PARTY -- PITCH BLACKNESS, THICK AND HEAVY, SOON ENGULFS THEM. THE SURROUNDING AIR BECOMES THIN AND STRANGULATING... ONE BY ONE THOUSANDS OF TORCHES WAVER, THEN GUTTER AND ARE EXTINGUISHED...
CEREBUS?
CEREBUS I'M SCARED.
FAR AWAY, ON THE WESTERN HORIZON, AN EERIE BLUE RADIANCE ENVELOPS THE EARTH, THE GROUND RISING AND FALLING LIKE WAVES ON A STORM-TOSSED OCEAN ...
FISSURES APPEAR FROM WHICH ISSUE PLUMES OF STEAM AND LEAPING TENDRILS OF ELECTRICAL FORCE -- HUGE SECTIONS OF EARTH BREAK FREE AND ARE HURLED SKYWARD ...
THE GROUND BENEATH THEIR FEET BEGINS TO VIBRATE AND HEAVE AS THE WEIRD LATTICEWORK OF RADIANCE EXPANDS...
AND BEGINS CRAWLING EASTWARD TOWARDS THEM...

SEDRA...
TEST
TRICK
TEST
TRICK
TEST
FSSSSS
PASS
FAIL
PASS
FAIL
PASS
FAIL
PASS
FAIL
PASS
FAIL
PASS
FAIL
PASS
FAIL

GOOD
BAD
GOOD
BAD
GOOD
BAD
GOOD
BAD
GOOD
BAD
GOOD
SEDRA
NOW, Y'ALL JUST SIT RIGHT THAR, SON ...
AND I'LL BE BACK SHORTLY WITH YOUR uh--
MORAL PURITY CLINICIAN...
PLEASE
TELL ME WHAT YOU WANT

QUEEN TAKES KING'S BISHOP'S PAWN.
CHECKMATE.

IN THE DISTANCE, A FORM BEGINS TO RISE AT THE EPICENTRE OF THE MASSIVE, HEAVING, WRENCHING CONVULSION...
I MUST GO AND SEE CIRIN...
HI...
FLESH
SPIRIT
FLESH
SPIRIT
FLESH
SPIRIT
FLESH
SPIRIT
FLESH
I AM BLOSSOM

SO-- YOU HAVE TO TAKE OFF WHOLE LEOTARD WHEN NATURE CALL?
TONS OF EARTH ARE DISPLACED, RISING, CRUMBLING, HURLED SKYWARD, THE SURROUNDING AIR THINS RAPIDLY, GROWING HOT AND STIFLING...
HUH?
uh NO·· THE TONGUE DESIGN HAS CONCEALED SNAP-FASTENERS HOLDING IT ON...

OOHHHH...
"SNAP FASTENERS"
YOU VERY CLEVER FELLOW, YES?
SNAP SNAP SNAP
PUFF PUFF PUFF
A GIANT HAND EMERGES AS THE SWELLING MOUND OF EARTH RISES HIGHER AND STILL HIGHER...

KNOCK
KNOCK
GREAT CIRIN?
YOU'VE BEEN SMOKING AGAIN...
MMM
YOU VERY BIG MAN
VERY STRONG ...
MMMMMM...
FEEL GOOD.
FEEL NICE.
PART OF MY ADAMANTIUM SKELETON...
GETTING HARDER STRONGER
CHANGING CHANGING INTO...

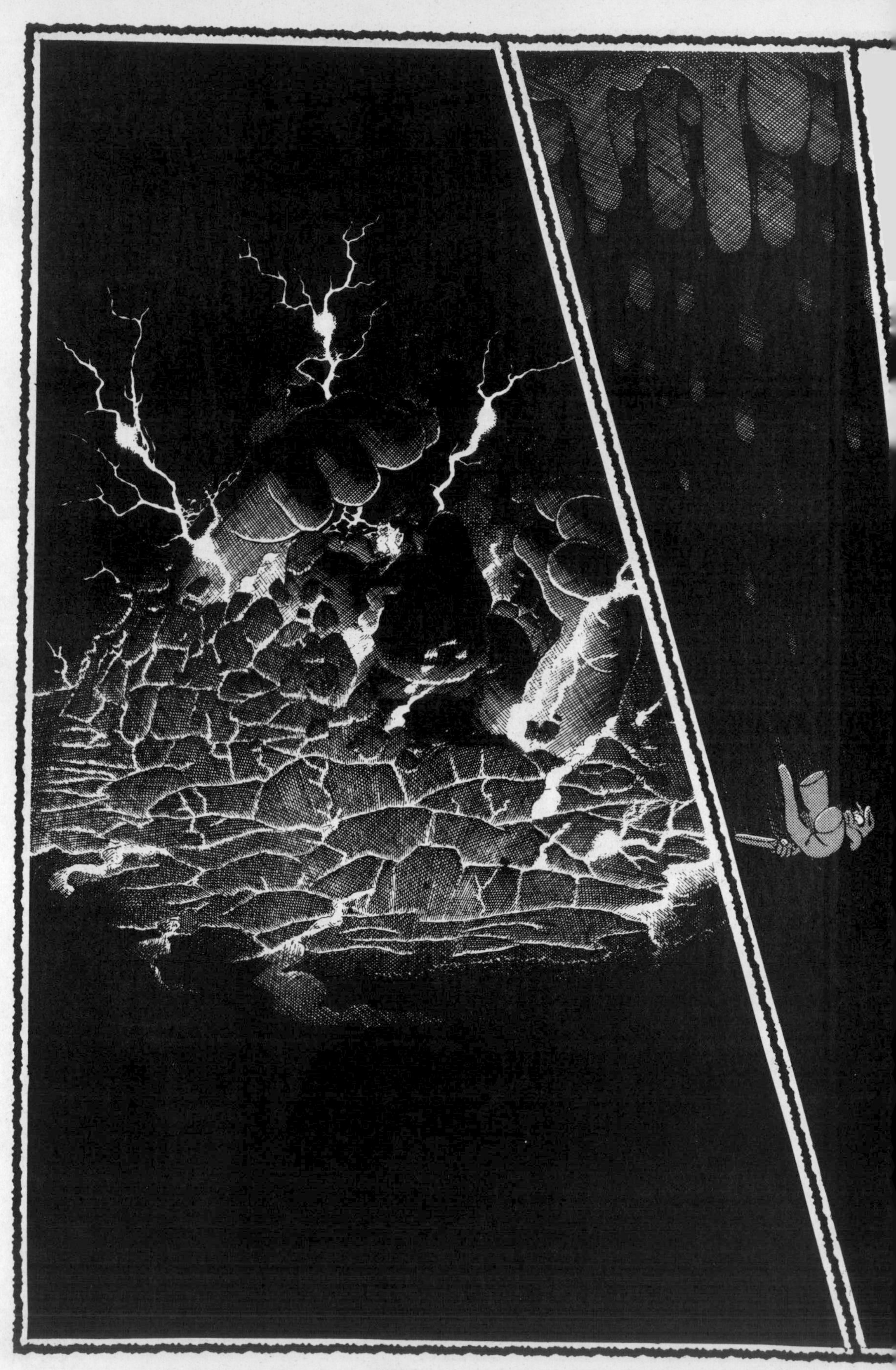

FUCK ME
LOBOROACH
FRAG!
FRAG OFF!
MOTHER FRAGGER!
FRAG YOU!
OH!
FRAG ME
OH!
FRAG ME
OH!
OH!
FRAG ME
CABLE ROACH
OH!
OH!
OH!
OHHHHHHHH
VENOMROACH
TINY HEAD.
TINY.
MY VENOM
STARTING TO... TO...
MR. HAMMOND AWAKENED ME A SHORT TIME AGO...
CONCERNING THE DIMENSIONS OF THE SPHERE
THEY'VE ...
YES?

OH
OH
OH
OH
OH
OH
TO--
OH!
GHOST ROACH
PUT VERY SIMPLY...
THE DIMENSIONS HAVE CHANGED... THE SPHERE IS NOW...
FORTY-SIX METERS ACROSS...

I KNOW GREAT CIRIN--
NO. THERE IS NO RATIONAL EXPLANATION
MR. HAMMOND ALSO INFORMS ME THAT SEVERAL TECHNICAL DIFFICULTIES HAVE APPARENTLY
RESOLVED THEMSELVES.
AGAIN
NO RATIONAL EXPLANATION

CORNERED, LORD JULIUS WITH NO PLACE TO RUN...
WHERE ARE YOUR BRAVE WITTICISMS NOW, EH?
FUNNY YOU SHOULD MENTION MY BRAVE WITTICISMS...
OWING TO THEIR REDUCED CIRCUMSTANCES THEY'VE OPTED FOR AN EARLY RETIREMENT TO A COZY LITTLE PROPERTY IN MY LOWER PELVIC REGION...
SPUT
FIZZ
SPUT
MMMMMMM

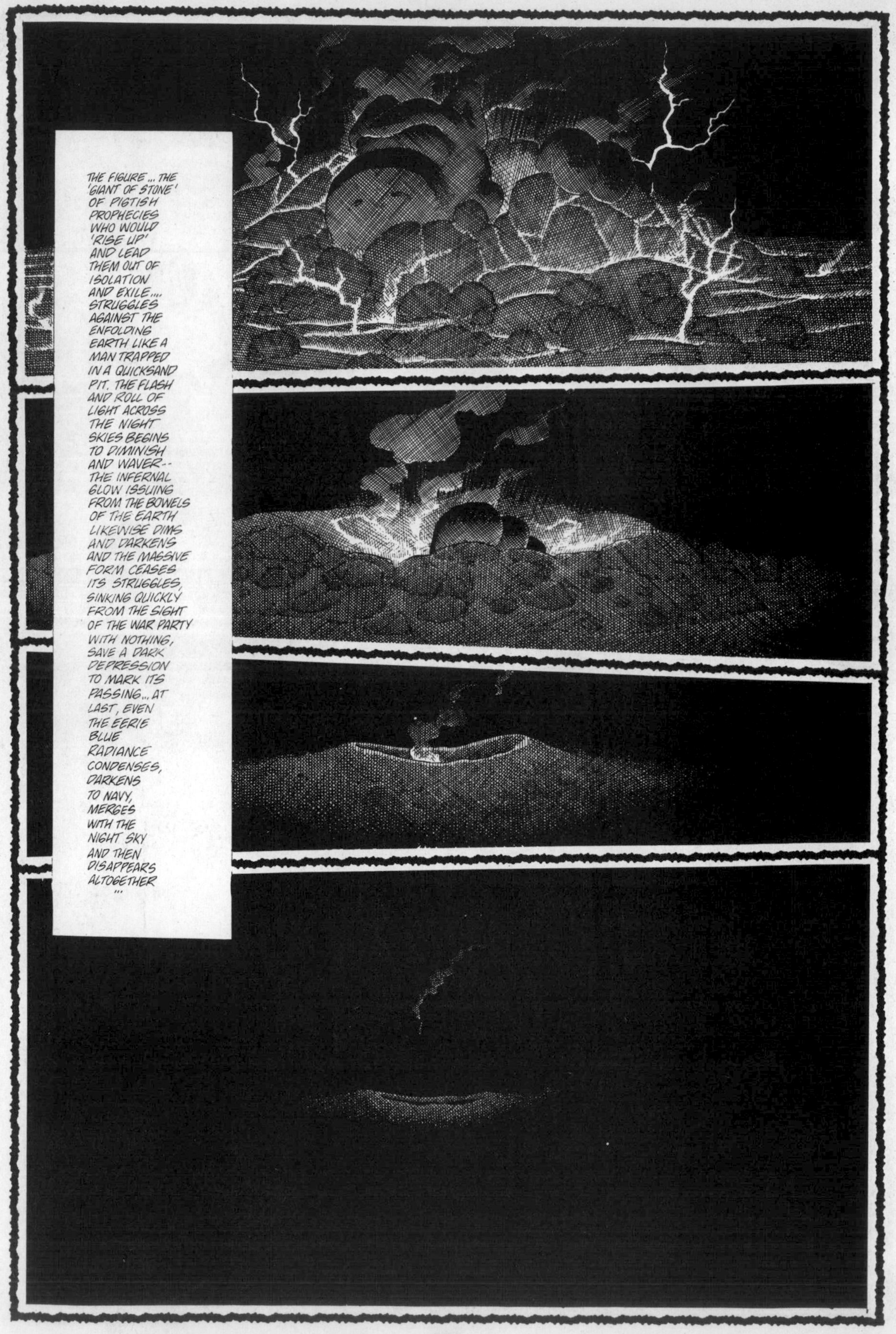
THE FIGURE ... THE 'GIANT OF STONE' OF PICTISH PROPHECIES WHO WOULD 'RISE UP' AND LEAD THEM OUT OF ISOLATION AND EXILE.... STRUGGLES AGAINST THE ENFOLDING EARTH LIKE A MAN TRAPPED IN A QUICKSAND PIT. THE FLASH AND ROLL OF LIGHT ACROSS THE NIGHT SKIES BEGINS TO DIMINISH AND WAVER-- THE INFERNAL GLOW ISSUING FROM THE BOWELS OF THE EARTH LIKEWISE DIMS AND DARKENS AND THE MASSIVE FORM CEASES ITS STRUGGLES, SINKING QUICKLY FROM THE SIGHT OF THE WAR PARTY WITH NOTHING, SAVE A DARK DEPRESSION TO MARK ITS PASSING... AT LAST, EVEN THE EERIE BLUE RADIANCE CONDENSES, DARKENS TO NAVY, MERGES WITH THE NIGHT SKY AND THEN DISAPPEARS ALTOGETHER ...

TUNG
CHK

I'VE
I'VE NEVER SHOT A WOMAN BEFORE
WELL, LET'S WAIT FOR THE AUTOPSY
YOUR RECORD MIGHT BE INTACT

GREAT CIRIN
I'VE DONE A GOOD DEAL OF THINKING ON THIS AND...

THANK YOU MRS. THATCHER THAT WILL BE ALL.

SO-- AH SAY--
SO! HOW DID YOUR heh-heh
"TEST" GO, SON

AT... LONG... LONG... LAST...
I... I HAVE FOUND
MY QUEEN.

WHOA THAR, SON--
QUEEN!!?
BUT-- BUT-- BUT--
SHE'S A--
A CLINICIAN-- YES-- I KNOW
A HUMBLE ORIGIN FOR SO MAJOR A CHARACTER TO BE SURE ...
EVEN SO...
THIS IS A CROSS-OVER--A TEAM-UP THAT'S BEEN WRITTEN IN THE STARS SINCE TIME BEGAN...

A MUST-HAVE DOUBLE-BAG MATCH MADE IN HEAVEN, OLD CHUM...
A HOT FIRST APPEARANCE WITH SOLID, LONG-TERM INVESTMENT POTENTIAL...
BUT-- er uh...
SHE'S uh SHE'S
NOT EXACTLY IN MINT CONDITION, SON...
IF YOU FOLLOW MAH MEANING ...

THERE'S A FEW CREASES ON HER SPINE --SURE-- MINOR WEAR--A LITTLE FADING...
BUT THERE'S NO FLAKING NO PIECES MISSING...
SHE'S FINE... VERY FINE...
AND WITH MY LOVE AND TIME AND PATIENCE
SHE CAN BE RESTORED OLD FRIEND--
SHE CAN BE PRISTINE ONCE MORE ...

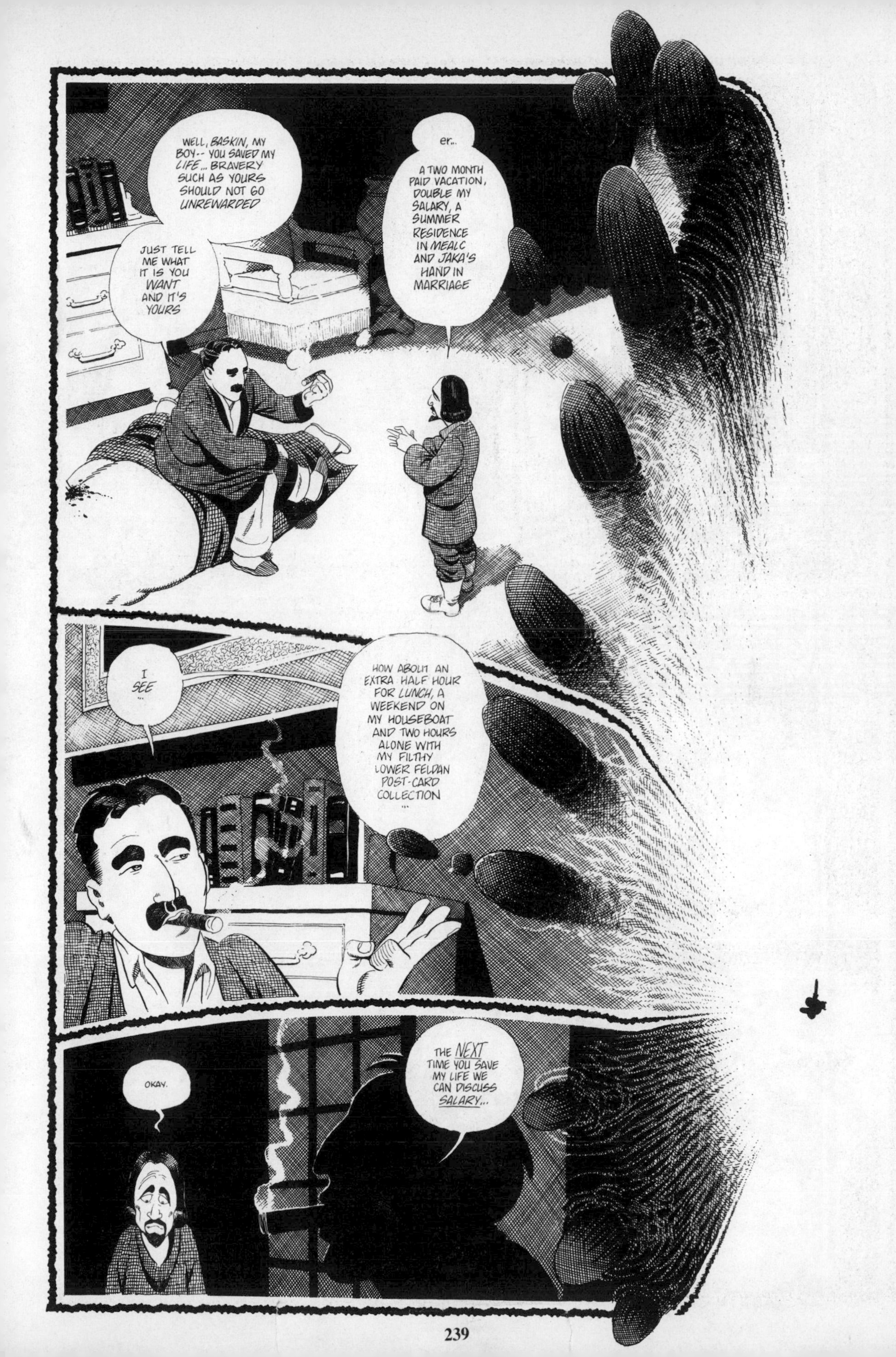
WELL, BASKIN, MY BOY-- YOU SAVED MY LIFE... BRAVERY SUCH AS YOURS SHOULD NOT GO UNREWARDED
JUST TELL ME WHAT IT IS YOU WANT AND IT'S YOURS
er..
A TWO MONTH PAID VACATION, DOUBLE MY SALARY, A SUMMER RESIDENCE IN MEALC AND JAKA'S HAND IN MARRIAGE
I SEE ...
HOW ABOUT AN EXTRA HALF HOUR FOR LUNCH, A WEEKEND ON MY HOUSEBOAT AND TWO HOURS ALONE WITH MY FILTHY LOWER FELDAN POST-CARD COLLECTION ...
OKAY.
THE NEXT TIME YOU SAVE MY LIFE WE CAN DISCUSS SALARY...

WE CAN'T JUST KEEP SWITCHING SIDES...
THE SACRED TEXTS ARE FULL OF FALSE PROPHETS...FAITH IS A CONSCIOUS CHOICE...
WE HAVE TO DECIDE RIGHT NOW...
WE HAVE TO PUT OUR FAITH IN TARIM
IF A PROPHET IS FALSE, THEN, WELL, TARIM WILL DESTROY HIM ...IT'S AS SIMPLE AS THAT
EVEN IF CEREBUS DOESN'T COME BACK WE ARE OBLIGED TO FOLLOW HIM
THE PUNISHEROACH IS HIS DISCIPLE...OR THAT'S WHAT HE SAYS...
I SAY HE IS...HE HAS TO BE...LOOK WHAT HE DID TO THE CIRINISTS ...
... THE SACRED TEXTS SAY THAT FALSE PROPHETS MUST BE BROUGHT DOWN
THEY DON'T SAY TARIM WILL DO THE JOB FOR US

HOW MANY TIMES IN HISTORY HAS IEST BEEN INVADED?
'GOING ALONG' IS THE ONLY ANSWER.
WHEN THE ORTHODOX TARIMITES ARE IN CONTROL, I'M AN ORTHODOX TARIMITE
WHEN THE CIRINISTS ARE IN CONTROL I'M A CIRINIST
IF CEREBUS COMES BACK?
"ALL HAIL MOST HOLY"
FOR NOW?
FOR NOW I'M A COCKROACHITE OR A COCK-ROACHIAN OR WHATEVER WE'RE CALLED
IT'S ALL AN ILLUSION-- THE ROACH. CIRIN. CEREBUS
ALL OF IT...
TIME AFTER TIME THE ONLY BELIEF THAT PROVES TO BE CORRECT--
ILLUSIONISM.
I LIKE THE CIRINISTS-- THE STREETS ARE SAFER NOW
I HOPE THEY KILL ROACH-MAN SO WE CAN ALL BE SAFE AGAIN AND STOP WORRYING ...
NO NO LISTEN
IT'S REALLY ALL BEEN SAID BEFORE ...AND NOTHING EVER CHANGES
I JUST GET TIRED OF ARGUING ...
BUILDING AUTOMATIC CROSS-BOWS, HARVESTING CROPS ...
ALL THE SAME TO ME...
JUST TELL ME WHERE TO REPORT FOR WORK IN THE MORNING

PRISTINE?
SON--AH DON'T LIKE TO RAIN ON ANYBODY'S PARADE...
EXCUSE ME D'EITHER O'YOU BOZOS KNOW WHAR I'D FIND "BLOSSOM" ?
BLOSSOM? CERTAINLY... SECOND FLOOR ON THE RIGHT! FRIEND OF HERS ARE YOU?
BELCH
PLEASURE TO MEET YOU, SIR A PLEASURE... SHE'S A WONDERFUL LADY...
I'VE BEEN LOOKING FOR SOMEONE LIKE HER FOR YEARS...
SHE GITS TH' JOB DONE ALL RIGHT
SON-- LISTEN-- AH SAY LISTEN TO ME
HAVE YOU EVER SEEN A MORE BEAUTIFUL SUNRISE... ?

THE RESULT IS IMMEDIATE AND EXPLOSIVE -- IN THE PITCH DARKNESS, FIGHTING ERUPTS ONCE MORE -- SENSELESS, BLIND AND BLOODY -- THE "GIANT OF STONE" WAS GONE! THOSE WHO HAD DOUBTED THE ADVENTURERS BLAMED THEM NOW FOR THE TRAGEDY WHICH HAD BEFALLEN THEIR PEOPLE
THE ADVENTURERS BLAMED THE DOUBTERS AND THEIR LACK OF FAITH FOR THE DEATH OF THEIR MOST SACRED DREAM. SOON, IT IS ENDED AND AS THE SUN APPEARS ON THE HORIZON, A MERE HANDFUL OF THE ORIGINAL WAR PARTY REMAINS, STAGGERING AIMLESSLY THROUGH A SEA OF PIGT CORPSES

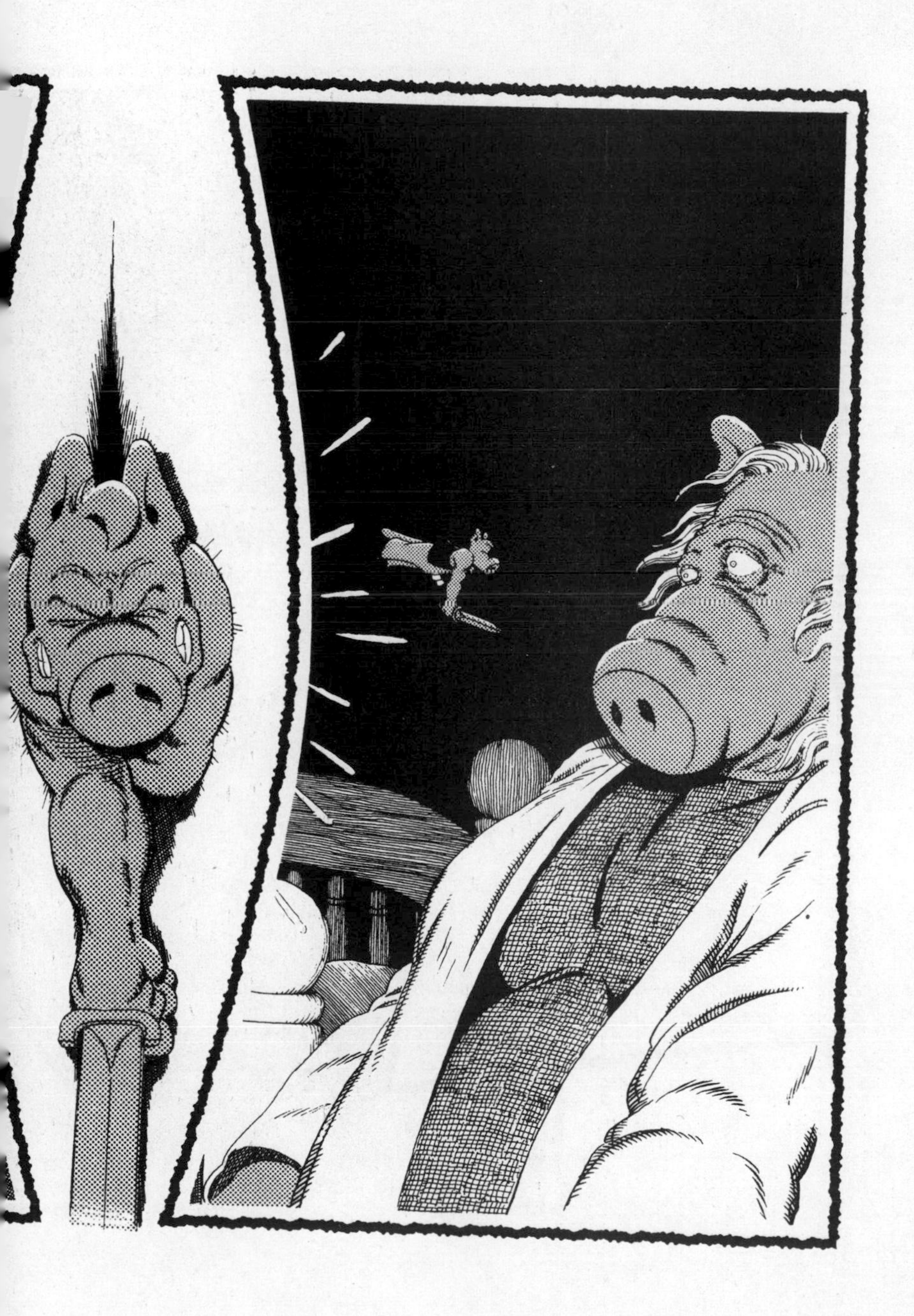

POIT!